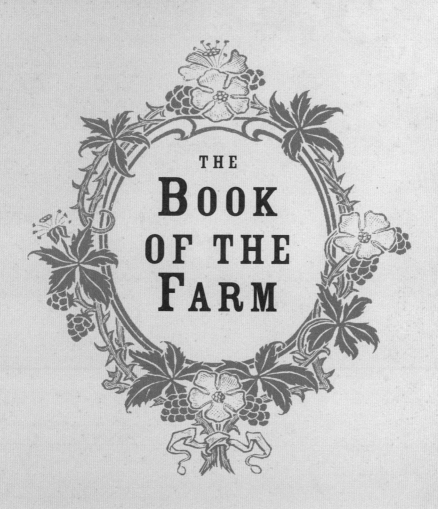

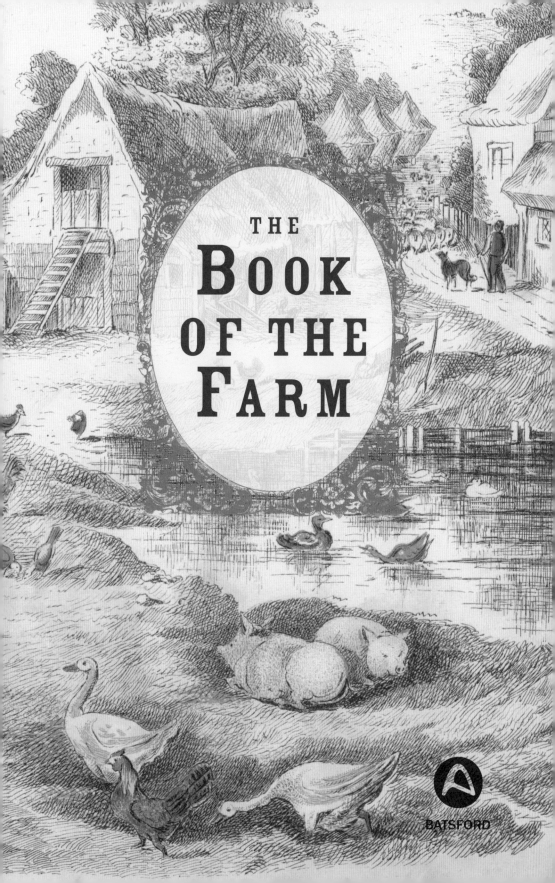

To my darling Libby

First published in the United Kingdom
in 2011 by
Batsford
10 Southcombe Street
London W14 0RA

An imprint of Anova Books Company Ltd

Copyright © Batsford 2011
Introduction and revised text © Alex Langlands 2011

The moral rights of the author have been asserted.

All rights reserved. No part of this publication
may be reproduced, stored in a retrieval system,
or transmitted in any form or by any means,
electronic, mechanical, photocopying, recording
or otherwise, without the prior written permission
of the copyright owner.

ISBN: 978-1-906388-91-1

A CIP catalogue record for this book is available
from the British Library.

18 17 16 15 14 13 12
10 9 8 7 6 5 4 3 2

Design by Lee-May Lim
Reproduction by
Mission Productions,
Hong Kong
Printed by Toppan
Leefung, China

This book can be
ordered direct from
the publisher at
www.anovabooks.co.uk,
or try your local
bookshop.

CONTENTS

Introduction 6

THE PERSONS WHO LABOUR ON THE FARM 18

WINTER 24
Introduction 26
Soils and subsoils 28
Planting of thorn hedges 31
The plough 34
Draining 42
Yoking and harnessing the plough,
    and swing trees 45
Drawing and storing turnips, cabbage,
    carrots and parsnips 50
Driving and slaughtering sheep 60
Rearing and feeding cattle on turnips
    in winter 68
Driving and slaughtering cattle 72
Treatment of farm horses in winter 78
Fattening, driving and slaughtering
    swine 82
The treatment of fowl in winter 86
Thrashing and winnowing grain,
    and the thrashing machine 92
Straw 102
    The forming of dunghills 104

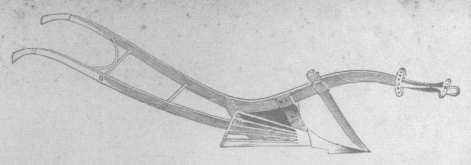

SPRING 106
Introduction 108
Cows calving, and calves 110
Sowing spring wheat 118
Sowing oat seed 124
Sowing barley seed 126
Seed germination 127
Sowing grass seed 130
Sowing flax and hemp 134
Sowing beans 136
Switching, pruning and water
    tabling thorn hedges 139
The lambing of ewes 146
Training and working the
    shepherd's dog 154
Turning dunghills and composts 158
Planting potatoes 162
Breaking in young draught horses 168
Sows farrowing or littering 172
The hatching of fowls 176

SUMMER 186
Introduction 188
Sowing turnips, mangelwurzel, rape,
    carrots and parsnips 190
Clearing stones, repairing fences,
    and the proper construction of
    field gates 198
Weaning calves and bulls, and grazing
    cattle till winter 203

Sheep washing, sheep shearing and
    weaning lambs 212
Making butter and cheese 222
Weeding corn, green crops,
    pastures and hedges, and
    casualties to plants 234
Haymaking 236
Summer fallowing and liming
    the soil 246
Building stone dykes 249

AUTUMN 252
Introduction 254
Pulling, steeping and drying flax
    and hemp 256
Harvesting rye, wheat, barley, oats, beans
    and pease 259
Carrying in and stacking wheat, barley,
    oats, beans and pease 272
Drafting ewes and gimmers,
    tupping ewes, and bathing and dipping
    sheep 280
Lifting and pitting potatoes 285
Sowing autumn wheat 290
Agricultural wheel carriages 294
Eggs 298

Conclusion 300
Index 301

INTRODUCTION

The middle decades of the nineteenth century represent the apogee of Queen Victoria's reign and the point at which Britain, globally, was at the height of its powers. *Britannia* was seen to be ruling the waves and in terms of its industry and economy, Britain was truly leading the world. At home, the towns and cities of the nation were growing at an unprecedented rate, with the ranks of the urban middle and working classes swelled to a population of unparalleled proportions. If the 'workshop of the world' was to keep pace with its global ambitions and see the threat of foreign imports kept at bay, it would need its larder stocked with basic foodstuffs to support its army of workers. This duty would fall, in the first instance, squarely upon the shoulders of British farmers. The optimism of the age, fuelled by the proven capital gains of large-scale industry and mass manufacture, was to spread its tentacles into agriculture. It gave fresh impetus to the existing developments that had begun to characterise what we now term the Agricultural Revolution, and the aim of the great farming magnates of the mid-nineteenth century was to turn the British landscape into a home-growing money-making machine.

It was in this climate that Henry Stephens' *Book of the Farm* was written. It was a time for British agriculture that has often been dubbed a 'Golden Age', when vast sums of

Henry Stephens

capital investment were set down to increase productivity and modernise farming practices. 'High Farming' – a term coined by James Caird, contemporary agriculturalist and writer – was the order of the day and a substantially remunerative farming was believed achievable through an intensive form of agriculture where high inputs led to high outputs. Nowhere better would the innovation and technical developments in farming have been observed than in the Agricultural Court at the Great Exhibition of 1851. Here the latest designs in agricultural implements vied for the attention of the finest entrepreneurial minds of the mid-Victorian Age. Not only could these new-fangled contraptions cut through the relentless back-breaking drudgery of an industry timelessly enslaved by the natural elements of weather, terrain and geology, but they could also increase productivity and, better still, generate that fundamental of any industry: profit.

WHAT IS *BOOK OF THE FARM* AND WHO WAS IT WRITTEN FOR?

The Book of the Farm is first and foremost a practical guide to farming in the Victorian period. Its focus is on the kinds of small details that make practical manuals so popular in the modern day and comprehensively guides the reader through the agricultural year of a 'mixed' farm; one that both breeds livestock for meat and dairy and grows crops for human and livestock consumption. From the types of farming that can be undertaken, to the breeds of livestock, the tools employed, the deployment of labour, the methods of sowing, feeding, weeding, harvesting and breeding, there is no aspect of farming that Stephens doesn't cover in minute detail. The resulting text, originally published in six volumes, is one that is, at times, verbose and confusing with details that seem, at first glance, irrelevant to the distant 21st-century reader. But to the budding mid-nineteenth century agriculturalist, this was the kind of detail that could shave down costs on every aspect of farming. It would have represented the best means with which to bypass the hard-earned experience and decades of knowledge gleaned from trial and error. It would also have included best farming practice elsewhere, detailing innovations and new ideas being written about in agricultural journals of the time and practised through experimentation on the leading farms of the nation.

In the Britain of the 1840s, agricultural education was a rather informal affair, with no clear national standard. Much of what was seen as best practice was passed down from either interested landowners or their farm bailiffs to their sons and younger members of the farming community. Similar arrangements are known, for all the country crafts and the agricultural skills of growing and breeding were very much kept in the family, where the farm cattleman, shepherd, waggoner and ploughman thought nothing of passing on their knowledge to their immediate or once-removed offspring. Put simply, you learnt 'on the job'. For the gentry, however, and the middling tenant farmers, the first stages of formal agricultural education were beginning to emerge. While it was not possible to take a degree in agriculture until the early twentieth century, the Agricultural College was founded in 1846 in Cirencester, Gloucestershire (later, in 1880, to become the 'Royal Agricultural College'). Alone as a college that set out to provide practical instruction in the embryonic field of agricultural science, there was still a sense in wider society of agriculture being a diversion for gentlemen perhaps rather than a business for farmers. There were some informal arrangements in existence whereby boys could take up residence at leading and progressive farms in the district and undertake a form of apprenticeship. Stephens himself benefited from a number of years of on-the-job learning at George Brown's farm at Whitsome Hill, Berwickshire.

Accessible and practical guides were, however, a feature of this period. Clark Hillyard's *Practical Farming and Grazing* (1836), for example, was a product of the popularity of a summary he'd written for the benefit of his son back in 1814. Expanding upon his notes, he was motivated by what he read in practical texts of the age and bemoaned of them that they were 'so verbose and theoretical that I soon laid them aside'. Another notable contemporary work was the *Manual of British Husbandry*, published in three volumes between 1834 and 1840, edited by J. French Burke under the auspices of the Society for the Diffusion of Useful Knowledge.

David Low's *Elements of Practical Agriculture* (1834) demonstrates the early links made between chemistry and farming. Low, like

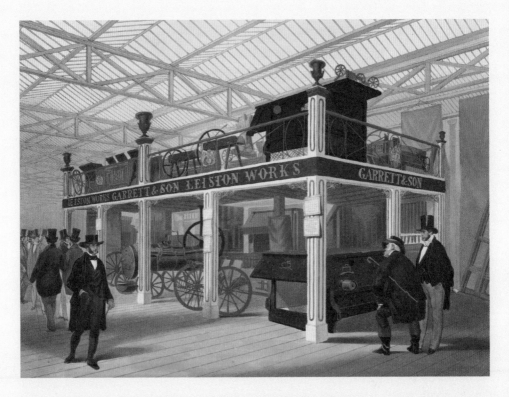

The Agricultural Court at the Great Exhibition, 1851

Stephens, had an early schooling in chemistry and had spent his formative years on his father's farm assisting in its day-to-day running. It is Humphrey Davy, however, in his 1813 publication *Elements of Agricultural Chemistry in a Course of Lectures* who should be credited, in Great Britain at least, with the first advocation of the application of the principles of chemistry to agriculture. Building on Davy's work, J. F. W. Johnston's *Elements of Agricultural Chemistry and Geology* was published in 1842, and in a climate where chemistry was increasingly being seen as key to the manufacture of fertilizers, the German chemist Justus Von Liebig's *Chemistry in its Application to Agriculture* was translated into English and published in 1840.

There is, of course, the task of turning the hard science of Davy and Liebig into practical application, and from there into hard cash. Low's book goes some way towards doing that, with his voice of experience from the fields providing balance. It is perhaps, with these more practically orientated texts, such as J. C. Morton's *Cyclopedia of*

Agriculture (1855) and J. C. Loudon's *Encyclopaedia of Agriculture* (1825) that Stephens' work sits most comfortably, although Stephens' departure from these was in his arrangement of information in seasonal order. In doing this, Stephens, more so than any agricultural writer of his day, was appealing to the newcomer. Thus, the enterprising young professional who wished to forge himself a career in the industry of agriculture, perhaps the son of a squire or a well-to-do tenant farmer, could be guided step-by-step through the agricultural year. The implementation of improvements and the introduction of new and innovative methods were very much the business of the land agent or farm manager, and *The Book of the Farm* would almost certainly have been one of the first acquisitions for the bookshelf, detailing best practice and highlighting the pitfalls as it does, in a clear and logically arranged seasonal order.

Who was Henry Stephens?

Henry Stephens was born in Bengal on 25th June 1795 to a surgeon, Andrew Stephens, of the East India Company. It was upon the death of his father in Calcutta on the 26th August 1806 that he was returned to the family home in Scotland and was from there sent to be educated in Dundee. From 1809 to 1811 he studied natural philosophy, mathematics and chemistry at the Dundee Academy under rector Thomas Duncan, and he went on to attend lectures in chemistry and agriculture at the University of Edinburgh where, incidentally, both Loudon and Low had also been educated.

It was on leaving Edinburgh and taking up an informal apprenticeship on George Brown's farm that he gained an understanding of all the processes entailed with mixed farming – from the day-to-day management of livestock to the seasonal field work required for the growing of crops. He worked in the dairy, the poultry house, the cattle byre, the sheep fold, the threshing barn and in the fields seeing and understanding first-hand the labour and method employed in the successful running of a farm. After several years working on Brown's farm, he travelled to Europe, where he explored Continental methods of farming, and upon his return to Scotland he took up the running of a large farm at Balmadies in the parish of Rescobie, Forfarshire. He is believed, through a programme of improvement, to have drastically improved the productivity of the farm at Balmadies and it is clear that many of the methods he advocates in *The Book of the Farm*, such as netting sheep on turnips in the field, a varied approach to drainage, and stock-proofing through well-maintained thorn-hedges and stone dykes, are given their first try-out at Balmadies.

Perhaps buoyed up by the success of his first years of farm management, Stephens began to write about his improvements. He published his initial findings in the *Quarterly Journal of Agriculture* (1829), which he went on to become editor of, and his taste for writing saw him take on the editorial of the *Transactions of the Highland and Agricultural Society*. But Stephens' position as a pioneer in the field of improved mixed farming was short-lived, for between 1830 and 1832 the failure of the chief East India Agency Houses at Calcutta through mismanaged business and over-speculated trade saw the collapse of his financial interests, and this was soon to prove calamitous to his farming ambitions.

Here there is an interesting parallel with another doyen of the 'High-Farming' age. John Joseph Mechi had made his fortune through a patent razor strop, and with his new-found wealth he bought a 130-acre farm at Tiptree in Essex, where he set about building the model farm. He was an ardent promoter, like Stephens, of investment in drainage, buildings, equipment and imported fertilisers as a means to raise fertility and increase production, and published his findings widely in talks to agricultural societies, in letters to the agricultural press and in his best-selling publication *How to Farm Profitably* (1859). However, it is clear that as his other business affairs started to take a downturn, so did his farming enterprise. In both Mechi and Stephens' cases it would appear that the 'best practice' that they so eloquently extolled was very much underpinned by external sources of funding – private capital – and their story is perhaps symbolic of the wider flaws in high farming that were so brutally exposed by the agricultural depression of the 1870s. Rising grain prices, a series of disastrous harvests and cheap imports had caused the collapse of the home-grown wheat market, and the ramifications had been catastrophic for many arable farmers in

A prize-winning Ayrshire cow owned by Mark J. Stewart Esq. of Southwick, Dumfries

Britain. Much of the capital that had been so enthusiastically invested in farm buildings, implements and increased soil fertility was never recovered, and many landowners and tenant farmers stared financial ruin in the face.

Fortunately for Stephens, his financial meltdown had been a generation earlier than the depression, and although the collapse of his investments meant that he had to give up Balmadies, he was able to turn his hand to writing for a still-optimistic agricultural community. Taking up residence in a suburb of Edinburgh, Stephens set about establishing himself as a full-time writer on all things agricultural. From 1842 to 1871 he published on numerous aspects of British farming and wrote *The Book of Farm Implements and Machines* (1858), *The Book of Farm Buildings* (1867), *Catechism of Practical Agriculture* (1855) and *A Manual of Practical Draining* (1846–48) but it is his first major work, *The Book of the Farm*, that is his most famous and most widely read. Beautifully illustrated by Gourlay Steell – more famous as Queen Victoria's animal painter and brother of John Steell the Scottish sculptor – *The Book of the Farm* proved an overriding success and numerous editions were published in Stephens' own lifetime (1842–4, 1849–51, and 1871) and beyond.

So why a new and revised edition of Book of the Farm?

Despite its popularity at the time, Stephens' *Book of the Farm* has received little recognition since the last editions were revised and published in the early twentieth century. It may very well be the case that for the general commentator on the changing fortunes of Britain's agricultural economy, Stephens' text is both impenetrable in its detail and somewhat myopic in its scope. Although there are occasional insightful observations on his part about the place that farming occupies in the national economy, teasing these out from pages and pages of technical and practical detail would require a depth of reading that would ultimately outweigh the fruits borne.

Yet it is precisely this level of detail that is the appeal for the archaeologist, historian and historical farmer like me. In the pre-industrial age, farming was the backbone of society, and the overwhelming majority of folk were rural dwellers concerned with the growing of food and the raising of stock to support the economy. Understanding this day-to-day existence on the land is therefore key to our understanding of pre-industrial societies. What Stephens provides us with is one of the first and most detailed commentaries on the technical issues – the tool use, the labour organisation, time taken, yields and management of resources. Although he was writing during a period of increased mechanisation, Stephens would regularly provide commentary on traditional methods in which the bulk of the work was carried out by labourers and draught animals. This is where the intrigue is for me. In these descriptions of processes conducted by hand can be found a window into agricultural practices that were, even in the mid-nineteenth century, hundreds of years old.

Of course, there are a myriad of other more contemporary sources available to the student of earlier agricultural history, but they come not without their problems of interpretation. For example, Thomas Tusser had, in the sixteenth century, provided us with a farming manual detailing *A Hundreth Good Pointes of Husbandrie* (first published in 1557). Yet, Tusser was a musician and poet first and a husbandman second, and as a consequence it often comes across that he is more at pains to achieve a rhyming couplet than he is to present an accurate

The month of August depicted in the *Julius Work Calendar*. Produced in Canterbury around the year 1020, this illustration depicts the bringing in of the harvest

portrayal of practical farming techniques. Virgil, the classical Roman poet, was a farmer's son, and his *Georgics* (literally meaning 'to farm' and dated to around 37BC) are rich in detail concerning the cultivation of crops and fruits and the raising of livestock in northern Italy. Again, however, the work has a motive beyond practical instruction and the rhetorical tone he sets creates tensions in purpose, often making it difficult to differentiate good from bad practice.

Early and later medieval manuscript illustrations and their depictions of farming life can be useful for the study of medieval farming. I have often found the *Julius Work Calendar*, an Anglo-Saxon manuscript produced in Canterbury Cathedral around 1020, instructive with regards to the various activities that might be carried out in the agricultural year. But caution is required, as the illustrator was undoubtedly a highly trained scribe rather than a skilled farmer, and the accuracy of each pictorial presentation should always be brought into question.

Similar problems arise from later medieval illuminated manuscripts such as the many surviving 'Books of Hours', popular devotional texts containing calendars of holy feasts, prayers and psalms. The various illustrations that feature in these texts are fraught with artistic licence and the biases of pious idealism alays need to be kept in mind. Having said this, however, the old adage 'a picture paints a thousand words' has some relevance here, as the detail in some of these illustrations is fascinating and I often find myself studying them intensely for insights into how the agricultural landscape was organised.

Of course, none of the authors or illustrators of the sources mentioned above set out primarily to give instruction in agricultural practices, and as such shouldn't be judged on these grounds. The student of historical farming who is interested in actual 'process' is left with the tricky business of piecing together oblique references from sweeping verse and tantalising detail from religiously loaded imagery to

form a greater understanding of the 'how to'. This, however, is where Stephens is so useful, as his first and only objective was to provide detailed instruction on method and practice. When we want to know exactly how, historically, people went about preparing the land, what methods they adopted to sow the seed, how they tended the crop in the

August in Da Costa Hours, where labourers take a welcome break from cutting cereals with sickles, illuminated by Simon Bening (1483/84–1561)

Reaping with the Hainault scythe. Developed in the Walloon region of Belgium in the sixteenth century, this method of reaping cereals was still being used well into the nineteenth century

field and how it was harvested under a range of conditions, it is to Stephens we turn.

To many, this obsession with the practical may seem a little dry and colourless, but it brings me to the crux of why I find *The Book of the Farm* so interesting. I am a great believer in making history work for itself and in the past being made as relevant as it can be for the present. While I love a good historical yarn about a king or queen, these tales are usually nothing more than stories of passing interest. Can we learn from them or is this history for amusement's sake? It is rare that the everyday people of the past have great narratives written about them, and we access their world through an understanding of the day-to-day tasks that made up their lives. Much of this, certainly in the pre-industrial age, was concerned with the provision and preparation of food – farming – and this is where I see *The Book of the Farm* as an historical document with surprising relevance for the now. It is my great belief that in these times of increasing uncertainty, in terms of how we feed, clothe and sustain ourselves, something of the methods of the past will find new audiences among society. We are already aware of the enormous carbon cost of global food production and our total reliance on non-renewable energy to sow, grow, harvest, process and transport our food. It is not just the stuff that is flown in from thousands of miles away, but that which is also grown in our immediate environment that is dependent on an ever-diminishing supply of fossil fuels. It may seem like a million years away now but for future generations – in maybe as little as fifty years' time – simply putting food on the kitchen table may prove extremely difficult. Confronted with this uncertainty I find myself flicking through the pages of *The Book of the Farm* and wondering whether some of the 'old ways' could still have a role to play. Of course, it would be naive of me to think that a reversion back to the methods of over one hundred years ago can simply replace the practices of today without a detrimental affect on yields and productivity. The situation is, unsurprisingly, much more complex than that. However, can we learn something from how once farms operated? Under 'closed' systems, fertility was kept in the ground by the dunging of livestock who were, in turn, fed by crops grown on the land, and in this we observe a cycle of life which supported those who lovingly tended and expertly managed the natural world around them. Alongside the wind, rain and sun, the source of energy was to be found in the muscular frames of the labourers, men and women, and in draught beasts, both horses and oxen. Yes, by the 1860s steam engines were playing an increasing role on the farm, and by the end of the century static oil engines made their first appearance, but before these it was people and horses that made food production on these islands tick. Should we today therefore, as individuals and groups, play a greater role in the production of the food we provide for ourselves, our families and our friends? Should the fields, meadows, woodlands, hedgerows and pastures that surround our towns and cities serve, as they once did, neighbouring markets and local outlets? Indeed, would it not be prudent for our local and regional communities to become less reliant on international markets for basic foodstuffs and thus more resilient to the whims and vagaries of a much less stable global economy?

The text I have selected from Henry Stephens' *Book of the Farm* leans heavily towards that which I find most interesting and that which excites the dreamer, like me, who believes that one day the old ways which served our ancestors so well will find a new momentum and experience something of a renaissance. Maybe, just maybe, there will come a time when more of us need to know how to harness a horse, how to set a plough, how to work a tilth and the depth to which the grain should be sown, and I hope that this revised edition of Henry Stephens' *Book of the Farm* will keep fresh in the mind of another generation the traditional skills of the Victorian farmer.

THE PERSONS WHO LABOUR ON THE FARM

The farmer

First, the duties of the *farmer*. It is his province to originate the entire system of management: to determine the period for commencing and pursuing every operation; to issue general orders of management to the steward, when there is one, and if there be none, to give minute instructions to the ploughmen for the performance of every separate field operation; to exercise a general superintendence over the fieldworkers; to observe the general behaviour of all; to see if the cattle are cared for; to ascertain the condition of all the crops; to guide the shepherd; to direct the hedger or labourer; to effect the sales of the surplus produce; to conduct the purchases conducive to the progressive improvement of the farm; to disburse the expenses of management; to pay the rent to the landlord, and to fulfil the obligations incumbent on him as a resident of the parish.

The bailiff

The duty of *steward*, or *grieve* as he is called in some parts of Scotland, and *bailiff* in England, consists of receiving general instructions from his master, the farmer, which he sees executed by the people under his charge. He exercises a direct control over the ploughmen and fieldworkers, and unreasonable disobedience on their part of

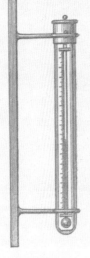

Fahrenheit thermometer, mounted

Funnel rain gauge

The gentleman farmer instructs one of his smock-clad labourers

his commands is reprehended as strongly by the farmer as if the affront had been offered to himself.

The farmer reveals to the steward alone the plans of his management; entrusts him with the keys of the corn barn, granaries and provision stores; delegates to him the power to act as his representative on the farm in his absence; and takes every opportunity of showing confidence in his integrity, truth and good behaviour.

The steward should always deliver the daily allowance of corn to the horses. He should, moreover, be the first person out of bed in the morning and the last in it at night.

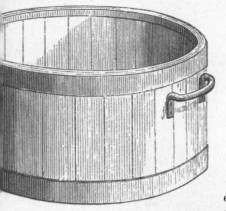

Imperial bushel

The ploughman

The duties of the *ploughman* are clearly defined. The principal duty is to take charge of a pair of horses, and work them at every kind of labour for which horses are employed on a farm. Horse labour on a farm is various. It is connected with the plough, the cart, sowing machines, the roller and the thrashing mill, when horse power is employed in the thrashing of corn; so that the knowledge of a ploughman should comprehend a variety of subjects. In the fulfilment of his duties, the ploughman has a long day's work to perform; for, besides expending the appointed hours in the fields with the horses, he must groom them before he goes to the field in the morning and after he returns from it in the evening, as well as at midday between the two periods of labour.

It is the duty of the ploughman to work his horses with discretion and good temper; not only for the sake of the horses, but that he may execute his work in a proper manner. It is also his duty to keep his horses comfortably clean.

The shepherd

His duty is to undertake the entire management of the sheep; and, when he bestows the pains he should on his flock, he has little leisure for any other work. His time is occupied from early dawn, when he should be among his flock before they rise from their lair, and during the whole day, to the evening, when they again lie down for the night. To inspect a large flock at least three times a day, over extensive bounds, implies a walking to fatigue. Besides this daily exercise, he has to attend to the feeding of the young sheep on turnips in winter, the lambing of the ewes in spring, the washing and shearing of the fleece in summer, and the bathing of the flock in autumn. And, over and above these major operations, there are the minor ones of weaning, milking, drafting and marking, at appointed times; not to omit the unwearied attention to be bestowed, for a time, on the whole flock, to evade the attacks of insects.

The only assistance which he depends upon in personally managing his flock is that of his faithful dog, whose sagacity in that respect is little inferior to his own.

A lad leads the heavy horses as the ploughman guides the plough

The cattleman

The services of the *cattleman* are most wanted at the steading in winter, when the cattle are all housed. He has the sole charge of them. It is his duty to clean out the cattle houses, and supply the cattle with food, fodder and litter, at appointed hours every day, and to make the food ready for them, should prepared food be given them.

Shepherd's house on wheels

A field labourer reaps a field of wheat with a sickle

Bullock holder

In summer and autumn, when the cows are at grass, it is his duty to bring them into the byre or to the gate of the field, as the case may be, to be milked at their appointed times; and it is also his duty to ascertain that the cattle in the fields are plentifully supplied with food and water. He should see the cows served by the bull in due time, and keep an account of the cows' reckonings of the time of calving. He should assist at the important process of calving.

An elderly person answers the purpose quite well, the labour being neither constant nor heavy, but well-timed and methodical. The cattleman ought to exercise much patience and good temper toward the objects of his charge, and a person in the decline of life is most likely to possess those qualities.

FIELDWORKERS

Fieldworkers are indispensable servants on every farm devoted to arable culture.

The duties of fieldworkers, as their very name implies, are to perform all the manual operations of the fields, as well as those with the smaller implements, which are not worked by horses. The *manual* operations consist chiefly of cutting and planting the sets of potatoes, gathering weeds, picking stones, collecting the potato crop, and filling drains with stones. The operations with the smaller implements are pulling turnips

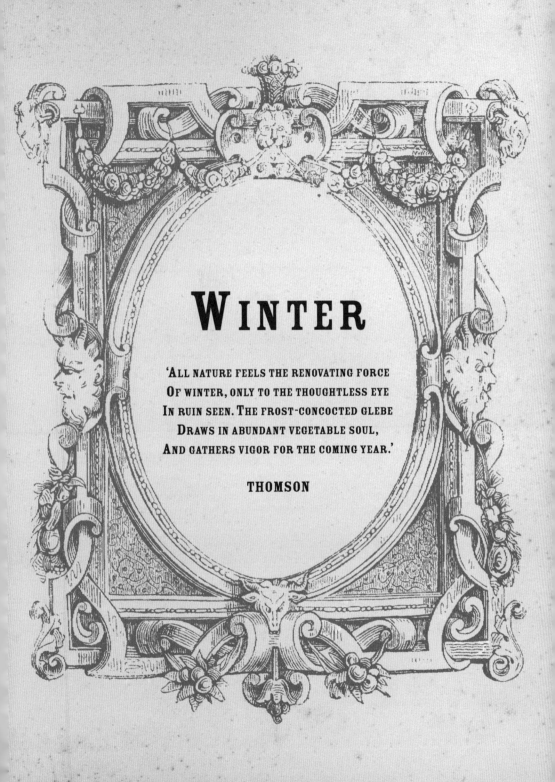

Winter

'All nature feels the renovating force
Of winter, only to the thoughtless eye
In ruin seen. The frost-concocted glebe
Draws in abundant vegetable soul,
And gathers vigor for the coming year.'

THOMSON

INTRODUCTION

The labours of the field in winter are confined to a few great operations. These are ploughing the soil in preparation of future crops and supplying food to the livestock. When the soil is naturally damp underneath, winter is the season selected for removing the damp by draining. Where fields are unenclosed and intended to be fenced with the thorn hedge, winter also is the season for performing the operation of planting it.

Almost the entire livestock of an arable farm is dependent on the hand of man for food in winter. It is this circumstance which, bringing the stock into the immediate presence of their owner, creates a stronger interest in their welfare than in any other season. The feeding of stock is so important a branch of farm business in winter that *it* regulates the time for prosecuting several other operations. It determines the quantity of turnips that should be carried from the field for the cattle in a given time, and causes the farmer to consider whether it would not be prudent to take advantage of the first few dry, fresh days to store up a quantity of them, to be held in reserve for the use of the stock during the storm that may be at the time portending. It also determines the quantity of straw that should be provided from the stack yard, in a given time, for the use of the animals; and upon this, again, depends the supply of grain that can be sent to the market in any given time.

All the stock in the farmstead in winter, that are not put to work, are placed under the care of the *cattleman*. The feeding of that portion of the sheep stock which are barren, on turnips in the field, is a process practised in winter. The flock of ewes roaming at large over the pastures requires attention in winter, especially in frosty weather, or when snow is on the ground, when they should be supplied with hay, or turnips when the former is not abundant. The preparation of grain for sale constitutes an important branch of winter farm business, and should be strictly superintended. A considerable portion of the labour of horses and men is occupied in carrying the grain to the market town and delivering it to the purchasers – a species of work which jades farm horses very much in bad weather. In hard frost, when the plough is laid to rest, or when the ground is covered with snow, and as soon as possible the farmyard manure is carried from the courts and deposited in a large heap, in a convenient spot near the gate of the field which is to be manured with it in the ensuing spring or summer.

The *weather* in winter is of the most precarious description and, being so, the farmer's skill to anticipate its changes in this season is severely put to the test. Seeing that all operations of the farm are so dependent on the weather, a familiar acquaintance with the local prognostics which indicate a change for the better or worse is incumbent on the farmer.

Winter is to the farmer the season of *domestic enjoyment*. The fatigues of the long summer day leave little leisure, and much less inclination, to tax the mind with study; but the long winter evening, after a day of bracing exercise, affords him a favourable opportunity, if he has the inclination at all, of partaking in social conversation, listening to instructive reading or hearing the delights of music. In short, I know of no class of people more capable of enjoying a winter's evening in a rational manner than the family of the country gentleman or the farmer.

Double horse cart

SOILS AND SUBSOILS

The leading characters of ordinary soils are derived from only two earths, *clay* and *sand*, and it is the greater or lesser admixture of these which stamps the peculiar character of a soil. The properties of either of these earths are even found to exist in what seems a purely calcareous [chalky] or purely vegetable soil. When either earth is mixed with decomposed vegetable matter, whether supplied naturally or artificially, the soil becomes a *loam*, the distinguishing character of which is derived from the predominating earth. Thus, there are *clay soils* and *sandy soils*, when either earth predominates; and when either is mixed with decomposed vegetable matter, they are then *clay loams* and *sandy loams*.

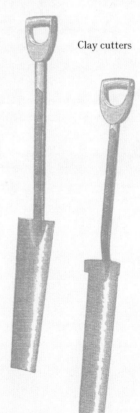

Clay cutters

A PURE CLAY SOIL

This has an unctuous feel in the hand, by which it can be kneaded into a smooth homogeneous mass, and retain any shape given to it. It is cold to the touch, and easily soils the hand and anything else that touches it. It cuts like soft cheese with the spade, and is then in an unfit state to be worked with the plough, or any other implement. A large strength of horses is thus required to work a clay land farm; for its workable state continues only for a short time, and it is the most obdurate of all soils to labour. But it is a powerful soil, its vegetation being luxuriant and its production great.

A SANDY CLAY SOIL

When a little *sand and gravel are mixed with clay*, its texture is very materially altered, but its productive powers are not improved. It does not easily ball in the hand. It renders water very muddy, and soils everything by adhering to it; and, on that account, never comes clean off the spade, except when much wetted with water. This kind of soil never occurs in deep masses, but is rather shallow; is not naturally favourable to vegetation, nor is it naturally prolific.

A CLAY LOAM

Clay loam – that is either of the clay soils mixed with a large proportion of naturally decomposed vegetable matter – constitutes a useful and valuable soil. It yields the largest proportion of the fine wheats raised in this country, occupying a larger surface of the country than the carse clay [literally meaning 'coarse' or 'rough']. It forms a lump by a squeeze of the hand, but soon crumbles down again. It is easily laboured, and may be so at any time after a day or two of dry weather. It is generally of some depth, forming an excellent soil for wheat, beans, Swedish turnips and red clover.

Clay soils are generally slow of bringing their crops to maturity, which in wet seasons they never arrive at; but in dry seasons they are always strong, and yield quantity rather than quality.

A PURE SANDY SOIL

A pure *sandy* soil is as easily recognized as one of pure clay. When wet, sandy soil feels firm underfoot, and then admits of a pretty whole furrow being laid over by the plough. It feels harsh and grating to the touch. In an ordinary state, it is well adapted to plants that have fusiform roots, such as the carrot and parsnip. Sandy soil generally occurs in deep masses, near the termination of the estuaries of large rivers, or along the sea shore, and is evidently a deposition from water.

A GRAVELLY SOIL

A *gravelly* soil consists of a large proportion of sand; but the greater part of its bulk is made up of small rounded fragments of rock brought together by the action of water. Gravelly deposits sometimes occupy a large extent of surface, and are of considerable depth. It can be easily laboured in any weather, and is not unpleasant to work, though the numerous small stones, which are seen in countless numbers upon the surface, render the holding of the plough rather unsteady. This soil is admirably adapted to plants having bulbs and tubers; and no kind of soil affords so dry and comfortable a lair to sheep on turnips, and on this account it is distinguished as 'turnip soil'.

SANDY AND GRAVELLY LOAMS

Sandy and *gravelly loams*, if not the most valuable, are certainly the most useful of all soils. They become neither too wet nor too dry in ordinary seasons, and are capable of growing every species of crop, in every variety of season, to considerable perfection. On this account, they are esteemed 'kindly soils'.

PLANTING OF THORN HEDGES

There are only two kinds of fences usually employed on farms, namely, thorn hedges and stone dykes. As winter is the proper season for planting, or running, as it is termed, thorn hedges, I shall here describe the process of planting the hedge.

The proper time for planting thorn hedges extends from the fall of the leaf, in autumn, to April, the latter period being late enough. The state of the ground usually chosen for the process is when in lea. If its line of direction is determined by existing fences (that is to say, if one side of a field only requires fencing), then the new fence should be made parallel with the old one that runs north or south, and it may take any convenient course, if its general direction is east and west. Should a field, or a number of fields, require laying off anew, the north and south fences should run due north and south, for the purpose of giving the ridges an equal advantage of the sun both forenoon and afternoon. To accomplish this parallelism a geometrical process must be gone through; and to perform that process with accuracy, certain instruments are required.

In the first place, three *poles* at least in number, of at least 8½ feet [2.6m] in length, should be provided. Three of such poles are required to determine a straight line, even on level ground; but if the ground is uneven, four or more are requisite. An *optical square* for setting off lines at right angles, or a *cross table*, for the same purpose, should also be provided. You should also have an imperial measuring chain, of 66 feet [20m] in length, which costs 13s, for measuring the breadth or length of the fields.

Being provided with these instruments, one line of fence is set off parallel to another in this way. Set off, in the first instance, at right angles, a given distance

Hand pick

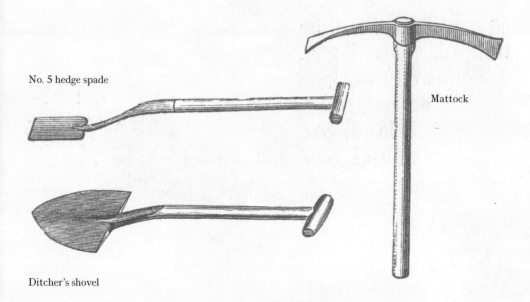

No. 5 hedge spade

Mattock

Ditcher's shovel

from near one end of the old thorn fence, if there be one, or of the ditch. At about 100 yards' [91.5m] distance, plant another pole in the same manner, and so on along the length of the fence from which the distances are set off. If there lie no fence to set off the distances from, then let a pole be set up perpendicularly in the line the new fence is intended to occupy and, at noon, on a clear day, observe the direction the shadow of the pole takes on level ground, and that is north and south; poles, at about 100 yards' [91.5m] distance, should be set up in the line of the shadow.

The next process is to plant it with thorns, and for this purpose certain instruments are required. 1. A strong *garden line* or *cord*. 2. A few *pointed pins of wood with hooked heads*. 3. A *wooden rule*, 6 feet [1.8m] in length, divided into feet and inches, having a piece of similar wood about 2 feet [60cm] in length, fastened at right angles to one end. 4. *No. 5 spades* (see above) are the most useful size for hedging. 5. A light *hand pick*. 6. An iron *tramp pick*. 7. A *ditcher's shovel* (see above).

Three men are the most convenient number to work together in running a hedge; and they should, of course, be all well acquainted with spade work. A *mattock* (shown above) for cutting will be required by each man, as well as a sharp *pruning knife*. The plant usually employed in this country, in the construction of a hedge, is the common hawthorn (shown right). 'On account of the stiffness of its branches,' says Withering, 'the sharpness of its thorns, its roots not spreading wide, and its capability of bearing the severest winters without injury, this plant is universally preferred for making hedges, whether to clip or grow at large.' Thorns ought never to be planted in a hedge till they have been transplanted at least two years from the seed bed.

In forming the *thorn bed*, raise a large, firm, deep spadeful of earth from the edge of the first rutted line of the hedge, and invert it along that line, with its rutted face toward the ditch. Having placed a few spadefuls in this manner, side by side, beat down their crowns with the back of the spade, paring down their united faces in the slope given to the first rut, and then slope their crowns with an inclination downward and backward from you, forming an inclined bed for the thorn plant to lie upon.

Push each plant firmly into the mould of the bed, with the cut part of the stem projecting not more than ½ of an inch [1.5cm] beyond the front of the thorn bed, and the root end lying away from the ditch, at distances varying from 6 to 9 inches [15 to 23cm]. The two assistants having finished laying the thorns, dig and shovel up with the spade all the black mould [loose soil] in the ditch, throwing it upon the roots and stems of the plants, until a sort of level bank of earth is formed over them. When the hedger has finished covering the plants with mould, and while the assistants are proceeding to clear all the mould from the ditch, he steps upon the top of the mound and, with his face toward the ditch, firmly compresses, with his feet, the mould above the plants, as far as they extend. By the time the compression is finished, all the mould will have been taken out of the ditch.

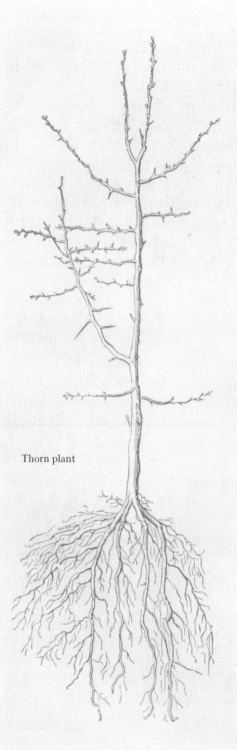

Thorn plant

THE PLOUGH

The plough serves the same purpose to the farmer as the spade to the gardener, both being used to *turn over* the soil, and the object of doing this is that this form of operation is the only means known of obtaining such a command over the soil as to render it friable and enclose manure within it, so that the seeds sown into it may grow into a crop of the greatest perfection. The *spade* is an implement so simple in construction, that there seems but one way of using it, whatever peculiarity of form it may receive, namely, that of pushing its mouth or blade into the ground with the foot, lifting up as much earth with it as it can carry, and then inverting it so completely as to put the upper part of the earth undermost. This operation, called *digging*, may be done in the most perfect manner; and any attempt at improving it, in so far as its uniformly favourable results are concerned, seems unnecessary. Hitherto it has only been used by the hand, no means having yet been devised to supply greater power than human strength to wield it. It is thus an instrument which is entirely under man's *personal* control.

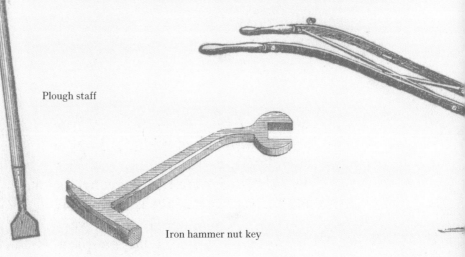

Plough staff

Iron hammer nut key

The effect attempted to be produced on the soil by the plough is an exact imitation of the work of the spade. From the circumstance, however, of the plough being too large and heavy an implement to be wielded by the hand, it is not so entirely under man's control as the spade. To wield it as it should be, he is obliged to call in the aid of horses, which, though not capable of wielding it personally, as man does the spade, can, nevertheless, through the means of appropriate appliances, such as harness, do so pretty effectually. It is not so simple a problem in practical mechanics, as it at first sight may appear, to construct a light, strong, durable, convenient instrument, which is easily moved about, and which, at the same time, though complex in its structure, operates by a simple action; and yet the modern plough (shown below) is an instrument possessing all these properties in an eminent degree.

A *good* ploughman will temper the irons, so there shall be no tendency in the plough to go too deep or too shallow into the ground, or make too wide or too narrow a furrow slice, or cause less or more draught to the horses, or less or more trouble to himself, than the nature of the work requires to be performed in the most proper manner. If he has a knowledge of the implement he works with – I mean, a good practical knowledge of it, for a knowledge of its principles is not requisite for his purpose – he will temper all the parts, so as to work the plough with great ease to himself and, at the same time, have plenty of leisure to guide his horses aright, and execute his work in a creditable manner.

A Ransome's plough

The action of the plough

The *coulter*, the *share* and the *mould board* being the principal active parts of the plough, and those which supply the chief characteristics to the implement, it may be useful to the farmer, as well as to the agricultural mechanic, to enter into a more minute descriptive detail of the nature and properties of these members.

The *coulter* is a simple bar which makes an incision through the soil, in the direction of the furrow slice that is to be raised. It is a remarkable fact that, in doing this, it neither increases nor decreases the resistance of the plough to any appreciable degree. Its sole use, therefore, is to cut a smooth edge in the slice which is to be raised, and an unbroken face for the land side of the plough to move against in its continued progress.

The *share* takes a more important and much more extensive part in the process. Its duty is very much akin to that of a spade. The share may be conceived as a spade wherein one of its angles has been cut off obliquely, leaving only a narrow point remaining, adapted to make the first impression on the slice. The share passes *under* the furrow slice, making a partial separation of it from the sole of the furrow, rising as the share progresses; the rise, however, being confined entirely to the *land side edge* of the slice – the furrow edge, as has been shown, remaining still in connection

View of the movement of the furrow slice

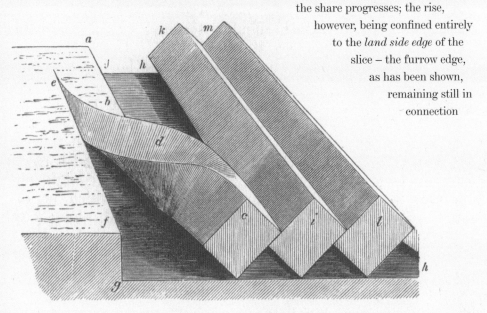

with the solid ground. The shield and back of the share are a continuation of the mould board; the latter, in its progress forward, receives the slice from the share and passes it onward or, more properly speaking, the plough passes under it.

The duty of the *mould board* is to transmit and deposit that slice in the best possible manner – the mould board is only a medium through which the slice is conveyed from the share to its destined position. To do this, however, in the most perfect manner, the mould board has to perform several highly important functions: first, the transmission of the slice; second, depositing it in the proper position; and third, performing both these operations with the least possible resistance.

In the case of the furrow slice movement, it appears as in the annexed perspective view (below left), where *a b is* the edge of the land as cut by the preceding furrow; *c d* the slice in the act of turning over; *e f*, the edge of the land from which the slice *c d* is being cut; *g k*, the sole of the furrow, and *i k* and *l m* are slices previously laid up.

The mode of ploughing ridges

The first process in ridging up land from the flat surface is called *feering* or *striking* the ridges. This is done by planting three or more of such poles (shown right), graduated into feet and half feet, as were recommended for setting off the lines of a fence, and which are used both for directing the plough employed to *feer* in straight lines, and for measuring off the breadth of the ridges into which the land is to be made up, from one side of the field to the other.

Feering pole

Land is *feered* for ridging in this way: Let *a b* (see page 39) represent the south and east fences of a field, of which let *x* be the head ridge or head hind, of the same width as that of the ridges, namely 15 feet [4.6m]. To mark off its width distinctly, let the plough pass in the direction of *r e*, with the furrow slice lying toward *x*. Do the same along the other headland, at the opposite end of the field. Then take a pole and measure off the width of a quarter of a ridge, *viz*. 3 feet 9 inches [115cm], from the ditch lip *a* to *c*, and plant a pole at *c*. With another pole set off the same distance from the ditch *a* to *d*, and plant it there. Then measure the same distance from the ditch at *e* to *f*, and at *f* look if

d has been placed in the line of fc; if not, shift the poles a little until they are all in a line. Make a mark on the ground with the foot, or set up the plough staff at f. Then plant a pole at g in the line of fdc. Before starting to feer, the ploughman measures off *1¼ ridges* – namely, 18 feet 9 inches [5.7m] – from f to k, and plants a pole at k. He then starts with the plough from f to d, where he stops with the pole standing between the horses' heads, or else pushed over by the tying of the horses. He then, with it, measures off, at right angles to fc, a hue equal to the breadth of 1¼ ridges, 18 feet 9 inches [5.7m], toward t until he comes to the line of kl, where he plants the pole. In like manner he proceeds from d to g, where he again stops, and measures off 1¼ ridges, 18 feet 9 inches [5.7m] breadth, from g toward v at a point in the line of kl, and plants the pole there. He then proceeds toward the other head ridge to the last pole c from g, and measures off 1¼ ridges, 18 feet 9 inches [5.7m], from c to l, and plants the pole at l. From l he looks toward k to see if the intermediate poles are in the line lk; if not, he shifts them to their proper points as he returns to the head ridge x along the furrow he had made in the line fc.

On coming down cf he obviates any deviation from the straight line that the plough may have made. In the line of fc the furrow slices of the feering have been omitted, to show you the setting of the poles. It is of much importance to the correct feering of the whole field to have those first two feerings, fc and kl, drawn correctly; and, to attain this end, it is proper to employ two persons in the doing of it – namely, the ploughman and the farm steward, or farmer himself. It is obvious that an error committed at the first feerings will be transmitted throughout the whole field. A very steady ploughman and a very steady pair of horses, both accustomed to feer, should only be entrusted with the feering of land. Horses accustomed to feer will walk up of their own accord to the pole standing before them. In like manner the ploughman proceeds to feer the fine kl, and so also the line op; but in all the feerings after the first, from f to k, the poles, of course, are set off to the exact breadth of the ridge determined on – in this case 15 feet [4.6m], such as from s to t, u to v, p to w, in the direction of the arrows. And the reason for setting off cl at so much a greater distance than lp or pw is that the half ridge ah may be ploughed up first and without delay, and that the rest of the ridges may be ploughed by half ridges.

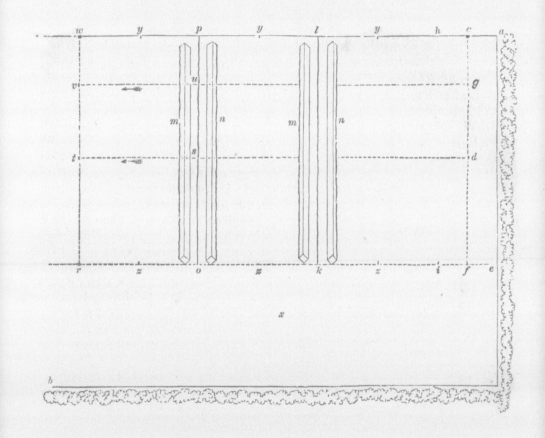

Mode of feering ridges

The half ridge $a\,h$ is, however, ploughed in a different manner from the rest; it is ploughed by going round the feering $f\,c$ until the open furrow comes to $a\,e$ on the one side and to $h\,i$ on the other. Then $h\,i$ constitutes the feering, along with $k\,l$, for ploughing the 2 half ridges $z\,i$ and $z\,k$, which, when done, the open furrow is left in the line $z\,y$, corresponding to the open furrow left in the line $e\,a$, and between which is embraced and finished the full ridge of 15 feet [4.6m] $e\,z$. The half ridges $z\,k$ and $z\,o$ are ploughed at the same time by another pair of horses, and the open furrow $z\,y$ is left between them, and the full ridge $z\,k\,z$ is then completed. In like manner the half ridges $z\,o$ and $z\,r$ are afterward ploughed by the same horses, and the open furrow $z\,y$ is left between them, and the full ridge $z\,o\,z$ is then completed.

As a means of securing perfect accuracy in measuring off the breadths of ridges at right angles to the feerings, lines at right angles to $f\,c$ should be set off across the field, from the cross table, and poles set at d and g, in the direction of $d\,t$ and $g\,v$, and a furrow made by the plough in each of these lines, before the breadths of the feerings are measured along them. Most people do not take the trouble of doing this, and a very careful ploughman renders it a precaution of not absolute necessity, but every proficient farmer will always do it, even at the sacrifice of a little time and some trouble, as a means of securing accuracy of work.

As the plough completes each feering, the furrow slices appear laid over as at m and n. While one ploughman proceeds in this manner to feer each ridge across the field, the other ploughmen commence the ploughing of the land into ridges; and, to afford a number of ploughmen space for beginning their work at the same time, the feering ploughman should be set to his work at least half a day in advance of the rest, or more, if the number of ploughs is great or the ridges to be feered long. In commencing to plough the ridges, each ploughman takes two feerings, and begins by laying the furrow slices of the feerings together, such as m and n, to form the crowns of the future ridges. In this way one ploughman lays together the furrow slices of $f\,c$ and $k\,l$, while another is doing the same with those of $o\,p$ and $r\,w$. I have already described how the half ridge $a\,h$ is ploughed, and stated that the rest of the ridges are ploughed in half ridges. The advantage of ploughing by half ridges is that the open furrows are thereby left exactly equidistant from the crowns; whereas, were the ridges ploughed by going round and round the crown of each ridge, one ridge might be made by one ploughman a little broader or narrower than the one on each side of it – that is, broader or narrower than the determinate breadth, which in this case is 15 feet [4.6m].

A ridge, $a\,a$, (below) consists of a crown b, two flanks c, two furrow brows d, and two open furrows $a\,a$. After laying the feering furrow slices to make the crowns of the ridges, such as at fc, $k\,l$, $o\,p$ and $r\,w$ (page 39), the plan to plough up ridges from the flat ground is to turn the horses toward you on the head ridges, until all the furrow slices between each feering are laid over until you reach the lines $y\,z$, which then become the open furrows. This method of ploughing is called gathering up, or gathering up from the flat, the disposition of whose furrows is shown (below) where $a\,a\,a$ embrace two whole ridges, on the right sides of which all the furrows lie one way, from a to b, reading from the right to the left; and on the left sides of which all the furrow slices lie in the opposite direction, from a to b, reading from the left to the right; and both sets of furrow slices meet in the crowns $b\,b\,b$.

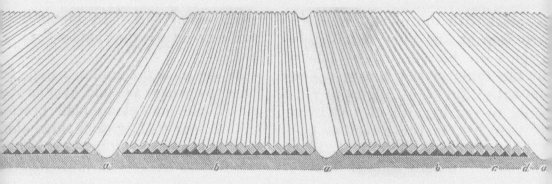

Gathering up from the flat

Draining

Draining may be defined as the art of rendering land not only so free of moisture that no superfluous water shall remain in it, but that no water shall remain in it so long as to injure or even retard the healthy growth of plants required for the use of man and beast. Indeed, my opinion is – and its conviction has been forced upon me by dint of long and extensive observation of the state of the agricultural soil over a large portion of the country – that this is the *true cause of most of the bad farming to be seen*, and that *not one farm* is to be found throughout the kingdom that would *not be much the better for draining*.

To the experienced eye, there is little difficulty in ascertaining the particular parts of fields which are more affected than others by superfluous water. They may be detected under whatever kind of crop the field may bear at the time; for the peculiar state of the crop in those parts, when compared with the others, assists in determining the point. There is a want of vigour in the plants; their colour is not of a healthy hue; their parts do not become sufficiently developed; the plants are evidently retarded in their progress to maturity; and the soil upon which they grow feels inelastic, or saddened under the tread of the foot. There is no mistaking these symptoms when once observed. They are exhibited more obviously by the grain and green crops, than by the sown grasses. In *old* pasture, the coarse, hard, uninviting appearance of the herbage is quite a sufficient indication of the moistened state of the soil.

Examinations of the soil and subsoil will tell you what kinds require deep draining, and what kinds may be treated with equal success under a different arrangement. There is more than one species of draining. There is one which draws off large bodies of water, collected from the discharge of springs in isolated portions of ground; and this is called *deep* or *under draining* (see illustration), because it intercepts the passage of water at a considerable depth under the surface of the ground. Another kind of draining absorbs, by means of numerous channels, the superabundant water spread over extensive pieces of

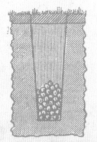    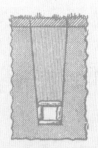

Methods of laying deep drains

ground under the surface, and has been called *surface draining*. This latter kind of draining subdivides itself into two varieties. One consists of small open channels formed on the surface of the ground in various directions for the ready use of water flowing upon the land, and this is literally *surface draining*. The other is effected by means of small drains constructed at small depths in the ground, at short distances from one another, and into which the water as it falls upon the surface finds its way by its own gravity through the loose soil, and by which it is discharged into a convenient receptacle.

There are *various ways of making small drains in grass*. One plan is to turn a furrow slice down the hill with the plough, and make the furrow afterward smooth and regular with the spade. When the grass is smooth and the soil pretty deep, this is an economical mode of making such drains, which have received the appellation of *sheep drains*. But where the grass is rough and strong, and swampy places numerous, the plough is apt to choke with long grass accumulating between the coulter and beam, and makes very rough work, and the horses are apt to overstrain themselves in the swampy ground. The lines of the drains should all be previously marked off with poles before the plough is used. A better, though more expensive, plan is to form them altogether with the spade.

Before beginning to break ground for thorough draining, it should be considered what quantity of water the drains will have to convey. The *drainage should be made to carry off the greatest quantity that falls, although it should occur only once in a lifetime*. In pursuing a system of drainage, every field should be thoroughly examined in regard to its state of wetness throughout the year, for that land is in a bad state which is soaking in winter, though it should be burnt up in summer; but the truth is, burning land requires draining as well as soaked land, because drains will supply moisture to burning land in summer, while they will render soaked land dry in winter.

The field having thus been fixed upon, the first consideration is the position of those drains that receive the water from the drains that are immediately supplied from the soil; and these are called *main drains*. No main drain should be put nearer than 5 yards [4.6m] to any tree or hedge, that may possibly push its roots toward it; but although the ditch of a hedge, whose roots lie in the opposite direction, merely receive the surface water from the field at the lowest end, it should not be converted into a main drain, that should be cut out of the solid ground, and not be nearer than 3 yards [2.7m] to the ditch lip; and the old ditch should be occupied by a small drain, and filled up with earth from the head ridge.

After having fixed the position of the main drains, and determined their levels and depths as here described, the next thing is the laying off of the *small drains*, which are so placed, or should be so constructed, as to have an easy descent toward the main drains into which they individually discharge their waters. They are usually cut in parallel lines down the declination of the ground; not that all the drains of the same field should be parallel to one another, but only those in the same plane, whatever number of different planes the field may consist of.

The first operation in breaking ground is to stretch the garden line for setting off the width of the top of the drain, 30 inches [76cm] – the drain being begun at the lowest part of the ground – and each division thus lined off consists of about four roods, or 24 yards [22m]. Three men are the most efficient number for carrying on the most expeditious cutting of drains. While the principal workman is ratting off the second side of the top of the drain with the common spade, the other two begin to dig and shovel out the mould earth, face to face, throwing it upon the lower and opposite side from the stones. The leading man then trims down the sides of the drain with his spade, and pulls out the remaining loose earth toward him with the scoop (see illustration) and thus finishes the bottom and sides in a neat, even, clean, square and workmanlike style.

Drain scoops

YOKING AND HARNESSING THE PLOUGH, AND SWING TREES

The first thing that will strike you is the extreme simplicity of the whole arrangement of the horses, harness, plough and man, impressing you with the satisfactory feeling that no part of it can go wrong, and affording you a happy illustration of a complicated arrangement performing complicated work by a simple action. On examining particulars, you will find the *collar* (right) around the horse's neck, serving as a padding to preserve his shoulders from injury while pressing forward to the draught.

Embracing a groove in the anterior part of the collar are the *haims* (overleaf), composed of two pieces of wood, curved toward their lower extremities, which are hooked and attached together by means of a small chain, and their upper extremities are held tight by means of a leather strap and buckle; and they are moreover provided on each side with an iron hook, to which the object of draught is attached.

The horse is yoked to the swing trees by light chains called *trace chains*, which are linked on one end to the hooks of the haims, and hooked at the

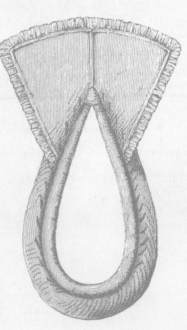

English draught horse collar

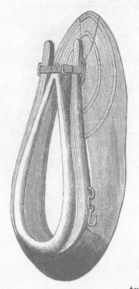

Scotch draught horse collar and haims

other into the eyes of the swing trees (shown right).

A *backhand* of leather put across the back, near the groins of the horse, supports the trace chains by means of simple hooks. The *bridle* has blinders, and while the horse is in draught, it is customary to hang the *bearing reins* over the tops of the haims. You observe there are two horses, the draught of the common plough requiring that number, which are yoked by the trace chains to the *swing trees*, which, on being hooked to the draught swivel of the bridle of the plough, enable the horses to exercise their united strength on that single point. The two horses are kept together either by a *leather strap*, buckled at each end to the bridle ring, or by a *short rein* of rope passed from the bridle ring to the shoulder of each horse, where it is fastened to the end of the trace chain with a knot. The ploughman guides the horses with *plough reins*, made of rein rope, which pass from both stilts to the bridle ring of each horse, along the outermost side of the horse, threading in their way a ring on the back band and sometimes another on the haims. The reins are looped at the end next to the ploughman, and conveniently placed for him under the ends of pieces of hard leather screwed to the foremost end of the helves [handles]; or small rings are sometimes put there to fasten the reins to.

Although the reins alone are sufficient to guide the horses in the direction they should go – and I have seen a ploughman both deaf and dumb manage a pair of horses with uncommon dexterity – yet the voice is a ready assistance to the hands, the intonations of which horses obey with celerity, and the modulations of which they understand, whether expressive of displeasure or otherwise.

The language addressed to horses varies as much as even the dialects are observed to do in different parts of the country. One word, *Wo*, to stop, seems, however, to be in general use. The motions required to be

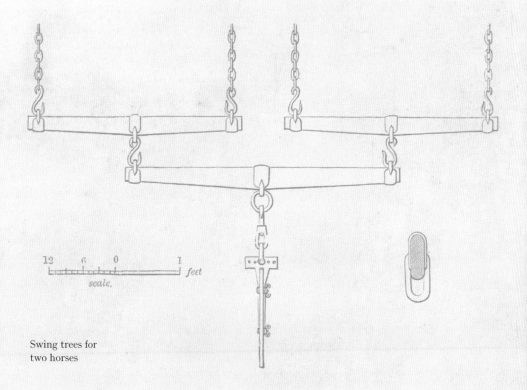

Swing trees for
two horses

performed by the horse at work are – to go forward, to go backward, to go from you and to come toward you; and the cessation of all these, namely, to stop or stand still.

To lessen or cease motion. The word *Wo* is the common one for a cessation of motion; and it is also used to make any motion slower; and it also means to be careful, or cautious, or not be afraid, when it is pronounced with some duration, such as *Wo o o*. In some parts, *Stand* has a similar signification; but, to stand without any movement at all, the word *Still* is there employed. In England, *Wo* is to stop.

To go forward. The name of the leader is usually pronounced, as also the well known *Chuck, Chuck*, made with the tongue at the side of the mouth, while impelling the breath.

To step backward. Back is the only word I can remember to have heard for this motion.

To come toward you. Hie is used in all the border counties of England and Scotland; *Hie here. Come ather*, are common in the midland counties of Scotland. In towns one hears frequently *Wynd* and *Vane*. In the west of England *Wo-e* is used.

To go from you. Hup is the counterpart to *Hie* in the southern counties, while *Hand off* is the language of the midland counties; and, in towns, *Haap* is used where *Wynd* is heard, and *Hip* bears a similar relation to *Vane*. In the west of England *Gee agen* is used. In all these cases, the speaker is supposed to be on what is called the *near side* of the horse – that is, on the horse's left side.

Shire stallion 'Honest Tom', bred by Mr Welcher of Watton, Norfolk, 1865

The swing tree or swing bars are those bars by which horses are yoked to the plough, harrows, and other implements. In the plough yoke, a set of swing trees consists of three (page 47) where *a* points at the bridle of the plough, *b b* the main swing tree attached immediately to the bridle, *c c* the furrow or offside little swing tree, and *d d* the land or nigh side little tree, arranged in the position in which they are employed in working. The length of the main tree, between the points of attachment for the small trees, is generally 3½ feet [1.1m], but this may be varied more or less; the length of the little trees is usually 3 feet [91cm] between the points of attachment of the trace chain, but this also is subject to variation.

Swing trees are for the most part made of wood, oak or ash being most generally used; but the former, if sound English oak, is by much the most durable, though good Scotch ash is the strongest, so long as it remains sound, but it is liable, by long exposure, to a species of decay resembling dry rot. As it is always of importance to know the why and wherefore of everything, I shall here point out how it may be known when a swing tree is of a proper degree of strength. A swing tree, when in the yoke, undergoes a strain similar in practice to that of a beam supported at both ends and loaded at the middle; and the strength of beams or of swing trees in this state is proportional to their breadths multiplied into the square of their depths and divided by their lengths. It is to be understood that the *depth* here expressed is that dimension of the swing tree that lies in the direction of the strain, or what in the language of the agricultural mechanic is more commonly called the *breadth* of the swing tree.

Drawing and Storing Turnips, Cabbage, Carrots and Parsnips

Turnip-lifter

The *treatment of livestock* receives early attention among the farm operations of winter; and whether they or land get the precedence depends entirely on the circumstance of the harvest having been completed late or early. Sheep always occupy the fields, according to the practice of this country, the only varieties of stock requiring accommodation in the steading in winter being cattle and horses. The horses consist chiefly of those employed in draught, which have their stable always at hand, and any young horses besides that are reared on the farm. Of the cattle, the cows are housed in the byre at night for some time before the rest of the cattle are brought into the steading.

By the time the cattle are ready to occupy the steading, *turnips* should be provided for them as their ordinary food, and the supply at all times should be sufficient; and it should be provided in the following way. The lambs of last spring, and the ewes which have been drafted from the flock as being too old or otherwise unfit to breed from any longer, are fed on turnips on the ground in winter, to be sold off fat in spring. The portion of the turnip ground allotted to sheep is prepared for their reception in a peculiar manner, by being *drawn* or *stripped*, that is, a certain proportion of the turnips is left on the ground for the use of the sheep, and the rest is carried away to the steading to be consumed by the cattle (see the chapter on cattle, page 68). The reason for stripping turnips is to supply food to the sheep in the most

convenient form, and, at the same time, enrich the ground for the succeeding crops by their dung.

The usual state in which turnips are stored is with their tops on, but the tails are generally taken away. The most cleanly state, however, for the turnips themselves, and the most nutritious for cattle, is to deprive them of both *tops* and *tails*. The tops and tails of turnips are easily removed by means of a very simple instrument (shown right).

The mode of using these instruments in the removal of the tops and tails of turnips is this. The fieldworker moves along between the two drills of turnips which are to be drawn, as from *a* (above) and pulling a turnip with the left hand by the top from either drill, holds the bulb in a horizontal position (see overleaf), over and between the drills *e* and *f* (above) and, with the knife, first takes off the root at *b* with a small stroke, then cuts off the top at *a*, between the turnip and the hand, with a sharper one, on which the turnip falls down into the heap.

Due care is requisite, on removing the tops and tails, that none of the bulb be cut by the instrument, as the juice of the turnip will exude through the incision. Of course, when turnips are to be consumed immediately, this precaution is less necessary; but the habit of slicing off a part or hacking the skin of the bulb indicates carelessness, and should be avoided at all times.

Method of pulling turnips in preparation for storing them

Turnip-trimming knife

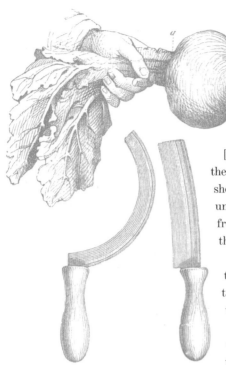

Mode of topping and tailing turnips

The topped and tailed turnips should be thrown into the carts by the hand, and not pricked by means of forks or graips [dung forks]; the cart should be placed alongside the drill near two or more heaps; and the carter should manage the horses and assist in the filling, until the turnips rise so high in the cart as to require from him a little adjustment in heaping, to prevent their falling off in the journey.

Dry weather should be chosen for the pulling of turnips, not only for the sake of cleanliness to the turnips themselves, but for the sake of the land, which should be cut up and poached by cart wheels and horses' feet as little as possible; because, when land is much cut up in carrying away turnips, sheep have a very uncomfortable lair, the ruts forming ready receptacles for water that are not soon emptied.

Should the weather prove unfavourable at the beginning of the season – that is, too wet or too frosty – there should no more turnips be pulled and carried than will suffice for the daily consumption of the cattle in the steading; but whenever the ground is dry and firm and the air fresh, no opportunity should be neglected except from other more important operations – such as the wheat seed – of storing as large a quantity as the time will permit, to be used when the weather proves interruptive to field operations.

The *storing* of turnips is very well done in this way: Let a piece of lea ground, convenient of access to carts, be chosen near the steading for the site of the store, and, if that be in an adjoining field, on a 15 foot [4.6m] ridge, so much the better, provided the ridge runs north and south. The cart with the topped and tailed turnips is backed to the spot of the ridge chosen to begin the store, and there emptied of its contents. The ridge being 15 feet [4.6m] wide, the store should not exceed 10 feet [3m] wide at the bottom, to allow a space of at least 2½ feet [75cm] on each side toward the open furrow of the ridge, for the fall and conveyance of water. The turnips may be piled up to the height of 4 feet [1.2m], but will not easily lie to 5 feet [1.5m] on that width of base. In this way, the store may be formed of any length; but it is more desirable to make two or three stores on adjoining ridges than a very long one on the same ridge, as its farthest end may be too far removed for using a wheelbarrow to remove the stored turnips. Assorted straw, that is, drawn out lengthwise, is put from 4 to 6 inches [10 to 15cm] thick above the turnips for thatch, and kept down by means of straw ropes, arranged lozenge-shaped, and fastened to pegs driven in a slanting direction in the ground, along the base of the straw, as may be distinctly seen in the figure below. Or a spading of earth, taken out of the furrow, may be placed upon the ends of the ropes to keep them down. The straw is not intended to keep out either rain or air – for both are requisite to preserve the turnips fresh – but to protect them from frost, which causes rottenness, and from dryness, which shrivels turnips. To avoid frost, the end, and not the side, of the store should be presented to the north, whence frost may be expected to come. If the ground chosen is so flat, and the open furrows are so nearly on a level with the ridges as that a dash of rain would overflow the bottom of the store, a furrow slice should, in that case, be taken out of the open furrows of the ridges with the plough, or a gaw cut made with the spade, and the earth used to keep down the ropes.

Best form of turnip-picker

Carrots may be stored exactly in the same manner, and so may *parsnips*. *Cabbages* are stored by being soughed into the soil, or hung up by the stems, with the heads downward, in a shed. As cabbages are very exhausting to the soil, the plants should be pulled up by the roots when they are gathered, and the stems not merely cut over with a hook or knife, because they will sprout again.

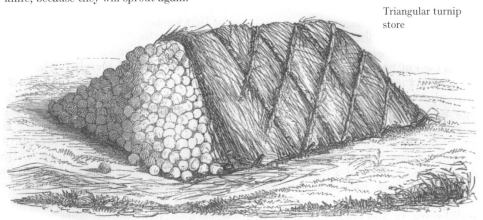

Triangular turnip store

Feeding sheep on turnips in winter

Having prepared room on the turnip land for the sheep intended to be fattened upon turnips, by removing the proportion of the crop in the manner described above, that is, by drawing two rows and leaving two rows alternately, and having prepared that part of the field to be first occupied by the sheep, which will afford them shelter in case of need, the first thing to be afterward done is to carry on carts the articles to the field requisite to form a temporary enclosure to confine the sheep within the ground allotted them. It is the duty of the shepherd to erect temporary enclosures, and as, in doing this, he requires but little assistance from other labourers, he bestows as much time daily upon it until finished as his avocations will allow.

There are two means usually employed to enclose sheep upon turnips, namely, by *hurdles* made of wood, and by *nets* of twine. Of these I shall first speak of the *hurdle* or *flake*. The mode of setting them is this; but in doing it, the shepherd requires the assistance of another person – a fieldworker will serve the purpose (see illustration below). The flakes are set down with the lower ends of their posts in the line of the intended fence. The first flake is then raised up by its upper rail, and the ends of the posts are sunk a little into the ground with a spade, to give them a firm hold. The second flake is then raised up and let into the ground in the same way, both being held in that position by the assistant. One end of a stay *f* is then placed between the flakes near the tops of their posts, and these and the stay are made fast together by the insertion, through the holes in them, of the peg *h*. The peg *i* is then inserted through near the bottom of the same posts. The flakes are then

Wooden hurdles or flakes set for confining sheep on turnips

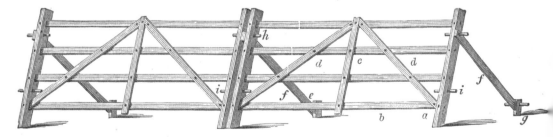

inclined backward away from the ground fenced, until their upper rail shall be 3 feet 9 inches [115cm] above the ground. The stake *e* is driven into the ground by the wooden mallet (shown right) at such a point as, where the stay *f* is stretched out from the flakes at the above inclination, that a peg shall fasten stake and stay together, as seen at *g*. After the first two flakes are thus set, the operation is easier for the next, as flake is raised after flake, and fastened to the last standing one in the manner described, until the entire line is completed.

Shepherd's hardwood mallet

The other method of enclosing sheep on turnips is with nets made of twine of the requisite strength. These nets having square meshes when stretched upon the stakes, usually extend to 50 yards [45.7m] in length, and stand 3½ feet [1.1m] in height. They are furnished with a rope along both sides passing through the outer meshes, which are called the 'top' and 'bottom' rope, as the position of either may be at the time. These ropes are wound round the stakes by a peculiar sort of knot called the 'shepherd's knot.'

When flakes or nets have been set round the first break, the ground may be considered in a proper state for the reception of sheep; and the ground should be so prepared before the grass fails, that the sheep to be fattened may not in any degree lose the condition they have acquired on the grass; for you should always bear in mind that it is much easier to improve the condition of lean sheep that have never been fatter than to regain the condition of those that have lost it. Much rather leave pastures a little rough, than risk the condition of sheep for the sake of eating it down. The rough pasture will be serviceable to the portion of the sheep stock that are not to be fattened, such as ewes in lamb and aged tups [rams]. Let sheep, therefore, intended to be fattened, be put on turnips as early as will maintain the condition they have acquired on the grass.

As the tops of white turnips are long and luxuriant at the commencement of the season, the first enclosure should be made smaller than those which succeed, that the sheep may not have too many tops at first on a change of food from grass to turnips, and which they will readily eat to excess, on account of their freshness and juiciness. Let the sheep fill themselves with turnips pretty well before taking them to the next break. The second break may be a little larger than the first, and the third may be of the proper size – that is, contain

Kirkwood's wire sheep-fodder rack

a week's consumption of food. When the tops wither in the course of the season, and one night of sharp frost may effect that, or after the sheep have been accustomed to the turnip, the danger is over. The danger to be apprehended is diarrhoea or severe looseness of the bowels, which is an unnatural state in regard to sheep, and they soon become emaciated by it; many sink under it, and none recover from such a relaxation of their system until after a considerable lapse of time.

Another precaution to be used on this head is to avoid putting sheep on turnips for the first time in the early part of the day when they are hungry. The afternoon, then, when they are full of grass, should be chosen to put the sheep on turnips, and they will immediately begin to pick the tops, but will not have time to injure themselves.

When sheep are on turnips they are invariably supplied with dry fodder, hay or straw – hay being the most nutritious, though most expensive; but sweet, fresh oat straw answers the purpose very well. The fodder is supplied to them in racks. There are various forms of straw racks for sheep – some being placed so high that sheep can with difficulty reach the fodder, and others are mounted high on wheels. The form represented (above) I have found convenient, containing as much straw at a time as should be given, admitting the straw easily into it, being easily moved about, of easy access to the sheep, and being so near the ground as to form an excellent shelter. Such a rack is easily moved about by two persons, and its position should be changed according to a change of wind indicative of storm.

It is the duty of the shepherd to supply these racks with fodder, and one or all of them may require replenishment daily. This he effects by carrying a bundle of fodder at any time he visits the sheep. When carts are removing turnips direct from the field, they carry out the bundles; but it is the duty of the shepherd to have them ready for the carters in the straw barn or hay house. For shelter alone the racks should be kept full of fodder. Fodder is required more at one time than another; in keen, sharp weather, the sheep eat it greedily, and when turnips are frozen they will have recourse to it to satisfy hunger, and after eating succulent tops they like dry fodder. In rainy, or in soft, muggy weather, sheep eat fodder with little relish; but it has been remarked that they eat it steadily and late, and seek shelter near the racks prior to a coming storm of wind and rain or snow; in fine weather, on the other hand, they select a lair in the more exposed part of their break.

I have constructed a bird's eye view of the manner in which a turnip field should be fitted up for sheep (see below). There are, in the first place, the turnips themselves *a*, of which half have been drawn by pulling two drills and leaving two alternately. The ground upon which they are growing is represented partly bare, because they are supposed to have been pulled up in the progress of the turnip cutter advancing from one side of the break to the other; and it constitutes the break.

Mode of occupying turnip fields with feeding sheep

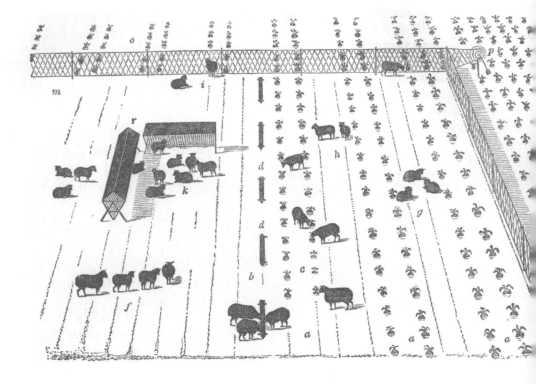

As matters are represented, the turnip slicer *b* is proceeding up beside the two drills *c*, and depositing the cut turnips into one of the small troughs *d*, out of another of which some of the sheep are eating, while others are helping themselves from the bulbs in the drills *c*. The sheep are represented scattered over the ground as they are usually seen, some following one another in a string *f* toward the place where their food is prepared for them, while others, *g*, are lying resting, regardless of food. Some, *h*, are standing, as if meditating what next to do, and others, *i*, examining the structure of the nets. Some nibble at the dry fodder in the racks *r*, while *k*, a group, lie under their shelter. Such are the usual occupations of sheep when they have abundance of food at their command. The nets *m* are represented as enclosing two sides of the break, the other two sides being supposed to be composed of the fences of the field, and not represented. The turnips *n*, to the right of the nets, appear undrawn, while those *o*, above the nets, are stripped, indicating that the progress of the breaks at this time is upward toward the top of the field.

Foot rot in sheep

On soft ground sheep are liable to be affected with *foot rot*, when on turnips. The first symptom is a slight lameness in one of the fore feet, then in both and, at length, the sheep is obliged to go down, and even creep on its knees, to get to its food. The hoof, in every case, first becomes softened, when it grows misshaped, occasioning an undue pressure on a particular part; this sets up inflammation, and causes a slight separation of the hoof from the coronet; then ulcers are formed below where the hoof is worn away, and then at length comes a discharge of fetid matter. If neglected, the hoof will slough off, and the whole foot rot off; which would be a distressing termination with even only one sheep, but the alarming thing is, that the whole flock may be similarly affected, and this circumstance has led to the belief that the disease is very contagious. There is, however, much difference of opinion among store farmers and shepherds on this point, though the opinion of contagion preponderates. For my part, I never believed it to be so, and there never would have been such a belief, had the disease been confined to a few sheep at a time; but though numbers are affected at one time, the fact can be explained from the circumstance of all the sheep being similarly situated; and as it is the nature of the situation which is the cause of the disease, the wonder is that any escape affection, rather than that so many are affected.

The first treatment for cure is to wash the foot clean with soap and water, then pare away all superfluous hoof, dressing the diseased surface with some caustic, the spirit of tar and blue vitriol being most in vogue, but Professor Dick recommends butter of antimony as the best; the affected part being bound round with a rag to prevent dirt getting into it again; and removing the sheep to harder ground, upon bore pasture, and there supplying them with cut turnips. The cure indicates the prevention of the disease, which is careful examination of every hoof before putting sheep upon red land, and paring away all extraneous horn; and should their turnips for the season be upon soft, moist ground, let them be entirely sliced, and let the sheep be confined upon a small enclosure at a time, and thus supersede the necessity of their walking almost at all upon it for food. I may mention that sheep accustomed to hard ground, when brought upon that which is comparatively much softer, are most liable to foot rot, and hence the necessity of frequent inspection of the hoof when on soft ground; and as some farms contain a large proportion of this state of land, frequent inspection should constitute a prominent duty of the shepherd.

Driving and Slaughtering Sheep

Sheep droving

It is proper that you should be made acquainted here with the *driving of sheep upon roads*, and the general practice of the *mutton trade*. Sheep should not begin their journey either when too full or too hungry; in the former state they are apt to purge on the road, in the latter they will lose strength at once. The sheep selected for market are the best conditioned at the time; and, to ascertain this, it is necessary to handle the whole lot, and shed the fattest from the rest; and this is best done about midday, before the sheep feed again in the afternoon. The selected ones are put into a field by themselves, where they remain until the time appointed them to start.

If there be rough pasture to give them, they should be allowed to use it, and get quit of some of the turnips in them. If there is no such pasture, a few cut turnips on a lea field will answer. Here all their hoofs should be carefully examined, and every unnecessary appendage removed, though the *firm* portion of the horn should not be touched. Every clotted piece of wool should also be removed with the shears. At the time the sheep should also be marked with *keil*, or *ruddle*, as it is called in England – the ochrey-red ironstone of mineralogists.

The *farmers' drover* may either be his shepherd, or a professional drover hired for the occasion. The shepherd knowing the flock makes their best drover, if he can be spared so long from home.

A drover of sheep should always be provided with a dog, as the numbers and nimbleness of sheep render it impossible for one man to guide a capricious flock along a road subject to many casualties; not a young dog, who is apt to work and bark a great deal more than

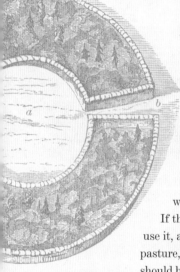

Circular stell

Opposite: A winter scene in which the farm sheepdog is sent out to bring the sheep in for the night

necessary, much to the annoyance of the sheep, but a knowing, cautious tyke [dog]. The drover should have a walking stick, a useful instrument at times in turning a sheep disposed to break off from the rest.

A shepherd's plaid he will find to afford comfortable protection to his body from cold and wet, while the mode in which it is worn leaves his limbs free for motion. He should carry provision with him, such as bread, meat, cheese or butter, that he may take luncheon or dinner quietly beside his flock while resting in a sequestered part of the road, and he may slake his thirst in the first brook or spring he finds, or purchase a bottle of ale at a roadside alehouse.

He should also have a good knife, by which to remove any portion of horn that may seem to annoy a sheep in its walk; and also a small bottle of a mixture of tobacco liquor and spirit of tar, with a little rag and twine, to enable him to smear and bandage a sheep's foot, so as it may endure the journey. He should be able to draw a little blood from a sheep in case of sickness. Should a sheep fail on the road, he should be able to dispose of it to the best advantage; or, becoming ill, he should be able to judge whether a drink of gruel, or a handful of common salt in warm water, may not recover it so as to

Lamb, ewe and ram

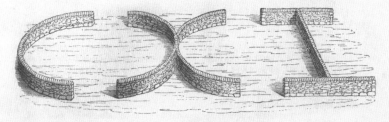
Ancient stells

proceed; but, rather than a lame or jaded sheep should spoil the appearance of the flock, it should be disposed of before the flock is presented in the market.

Assessing a 'ripe' sheep

Taking these data for your guide, you will be able to detect, by handling, the state of a sheep in its progress toward ripeness. A *ripe* sheep, however, is easily known by the *eye*, by the fullness exhibited in all the external parts of the particular animal.

A full-looking sheep need hardly be handled on the rump, for he would not seem so full unless fat had been first deposited there. A thin-looking sheep, on the other hand, should be handled on the rump, and if there be no fat there, it is useless handling the rest of the body, for assuredly there will not be so much as to deserve the name of fat. But between these two extremes of condition there is every variety to be met with; and on that account examination by the hand is the rule, by the eye alone the exception; but the hand is much assisted by the eye, whose acuteness detects deficiencies and redundancies at once.

In handling a sheep the points of the fingers are chiefly employed, and the accurate knowledge conveyed by them through practice of the true state of condition is truly surprising, and settles a conviction in the mind that some intimate relation exists between the external and internal state of an animal. And hence this practical maxim in the judging of stock of all kinds, that no animal will appear *ripe to the eye*, unless as much fat had previously been laid on in the inside as his constitutional habit will allow. The application of this rule is easy. Thus, when you find the rump nicked on handling, you may expect to find fat on the back; when you find the back nicked, you would expect the fat to have proceeded to the top of the shoulder and over the ribs; and when

Cheviot ram

you find the top of the shoulder nicked, you would expect to find fat on the underside of the belly. To ascertain its existence below, you will have to *turn him up*, as it is termed; that is, the sheep is set upon his rump with his back down and his hind feet pointing upward and outward. In this position you see whether the breast and thighs are filled up. Still, all these alone would not let you know the state of the inside of the sheep, which should, moreover, be looked for in the thickness of the flank; in the fullness of the breast, that is, the space in front from shoulder to shoulder toward the neck; in the stiffness and thickness of the root of the tail, and on the breadth of the back of the neck. All these latter parts, especially with the fullness of the inside of the thighs, indicate a fullness of fat in the inside; that is, largeness of the mass of fat on the kidneys, thickness of net, and thickness of layers between the abdominal muscles. Hence the whole object of feeding sheep on turnips seems to be to lay *fat* upon all the bundles of fleshy fibres, called muscles, which are capable of acquiring that substance; for as to bone and muscle, these increase in weight and extent independently of fat, and fat only increases their magnitude.

    I have spoken of the *turning up* of a fat sheep; it is done in this way. Standing on the near side of the sheep, that is, at its left side, put your left hand under its chin, and seize the wool there, if rough, or the skin, if otherwise; place your knees, still standing, against its ribs then, bowing forward a little, extend your right arm over the far loin of the sheep, and get a hold of its flank as far down as you can reach, and there seize a large and firm hold of wool and skin. By this, lift the sheep fairly off the ground, and turning its body toward you upon your left knee under its near ribs, place it upon its rump on the ground with its back to you, and its hind feet sticking up and away from you. This is an act which really requires strength, and if you cannot lift the sheep off the ground, you cannot turn it; but some people acquire a sleight in doing it, beyond their physical powers.

By-products of sheep

The sheep is one of the most useful, and therefore one of the most valuable, of our domestic animals; it not only supports our life by its nutritious flesh, but clothes our bodies with its comfortable wool. All writers on diet have agreed in describing mutton as the most valuable of the articles of human food.

But the products of sheep are not merely useful to man, they also promote his luxuries. The skin of sheep is made into leather and, when so manufactured with the fleece on, makes comfortable mats for the doors of our rooms, and rugs for our carriages. For this purpose the

Mode of cutting up a sheep:
1. Shoulder
2. Breast
3. Loin
4. Best-end neck
5. Scrag-end neck
6. Head
7. Loin (chump end)
8. Leg

'The utility of sheep to man in their employment during life and uses after death' 1845

best skins are selected, and such as are covered with the longest and most beautiful fleece. Tanned sheepskin is used in bookbinding. White sheepskin, which is not tanned, but so manufactured by a peculiar process, is used as aprons by many classes of artisans and, in

agriculture, as gloves in harvest; and, when cut into strips, as twine for sewing together the leathern coverings and stuffings of horse collars. Morocco leather is made of sheepskins as well as of goats, and the bright red colour is given to it by cochineal. Russian leather is also made of sheepskins, the peculiar odour of which repels insects from its vicinity, and resists the mould arising from damp – the odour being imparted to it in currying, by the empyreumatic oil of the bark of the birch tree. Besides soft leather, sheepskins are made into a fine, flexible, thin substance, known by the name of parchment; and, though the skins of all animals might be converted into writing materials, only those of the sheep and she goat are used for parchment. The finer quality of the substance called vellum is made of the skins of kids and still-born lambs, and for the manufacture of which the town of Strasburg has long been celebrated.

Mutton suet is used in the manufacture of common *candles*, with a proportion of ox tallow. Minced suet, subjected to the action of high pressure steam in a digester at 250°F or 260°F [121°C or 127°C], becomes so hard as to be sonorous when struck, whiter, and capable, when made into candles, of giving very superior light. Such candles, the late invention of the celebrated Guy Lussac, are manufactured solely from mutton suet. In the manufacture of some sorts of cords from the intestines of sheep, the outer peritoneal coat is taken off and manufactured into a thread to sew intestines, and make the cords of rackets and battledores. Future washings cleanse the guts, which are then twisted into different sized cords for various purposes. Some of the best known of those purposes are whip cords, hatters' cords for bow strings, clock makers' cord, bands for spinning wheels (which have now almost become obsolete) and fiddle and harp strings.

Rearing and Feeding Cattle on Turnips in Winter

The first thing to be done with the courts in the steading, before being taken possession of by the cattle, is to have them *littered plentifully with straw*. The first littering should be abundant, as a thin layer of straw upon the bare ground makes an uncomfortable bed; whereas a thick one is not only comfortable in itself, but the lower part of it acts as a drainer to the heap of manure above it. There is more of comfort for cattle involved in this little affair than most farmers seem to be aware of; for it

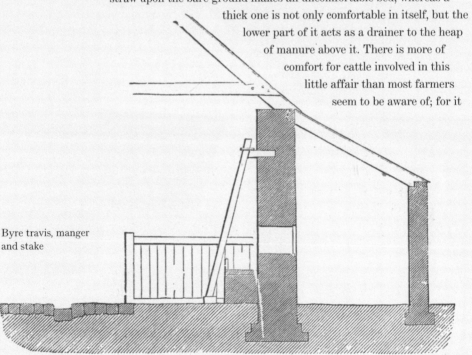

Byre travis, manger and stake

is obvious that the first layer of litter, when thin, will soon get trampled down, and in rainy weather soon become poached – that is, saturated with wet and pierced with holes by the cattle's feet – so that any *small* quantity of litter that is afterward laid upon it will but absorb the moisture below it, and never afford a *dry* lair to the cattle. On the other hand, when the first layer is thick, it is not poached even in wet weather, because it is with difficulty pierced through by feet, and it instantly drains the moisture that falls upon it, and of course keeps the bedding comparatively dry.

Each cow should always occupy the stall she has been accustomed to, and all will then go and come into their stalls without interfering with one another. They thus learn to become very quiet in the stall, both to the cattleman who feeds them, and the dairymaid who milks them. Each stall should have a manger (shown below right) elevated 20 inches [51cm] above the floor, lined with wood or stone, and having an edging of plank 8 inches [20cm] in depth, to keep in the food. There should be as much room behind the manger to the gutter as to allow the cow to lie at ease, whatever be her size, like a horse in a stall with a low hay rack. Each stall should have a travis board to separate it from the next.

Calves of the year should occupy the court. In such receptacles they are put together male and female, strong and weak, but having plenty of trough room around two of the walls, they can all be amply provided with food at the same time, without the danger of the stronger buffeting about the weaker. In the centre of the court stands the straw rack (right) and the troughs for the turnips are fitted (shown overleaf). There is a water trough in the same court, it being essential for young stock to have water at will, and especially when they do not get as many turnips as they can eat. When they do, cattle do not feel the want of water, the juice of the turnip supplying them with sufficient liquid.

Wooden straw rack for courts

And to begin with the cows. The first piece of work connected with the treatment of cows in winter, is to milk them at daybreak, which cannot be at a very early hour this season. On farms on which cows are bred, they are heavy in calf in winter; so most of them will be dry, and those still yielding milk, being the latest to calve, will give but a scanty supply. It is, therefore, not as *milk cows* that they are treated this season. I would, therefore, give them a mouthful of fresh oat straw, to prepare the stomach for the turnips. While amusing themselves with this fodder, the cattleman, whose duty it is to take charge of all the cattle in the steading in winter, cleans out the byre of its litter and dung with the graip (below right) and wheelbarrow (right) and spreads it equally over the court, and sweep the gutter and causeway clean with a birch or besom broom (shown on page 79). Having shut the byre door and left the half door into the court open for fresh air, the cattle man leaves the cows until he has supplied the fattening and young beasts with turnips which, having done, he returns to the cow byre, bringing litter straw with him, and gives them their allowance of turnips for the first meal.

Turnip trough for courts

The turnips should always be put into the troughs in a regular order, beginning at the same end of the byre, and finishing at the other, and after the turnips have been given, the cows should be permitted to eat them in quiet, for nothing irritates animals more than to be handled and worked about when feeding. The turnips consumed, and the stalls comfortably littered with straw, the cows will lie down and chew the cud until midday, when they should be turned into the court to enjoy the fresh air, lick themselves and one another, drink water from the trough and bask in the sun. They should go out a while every day, in all weathers, until they calve, except perhaps on a very cold, wet day.

Whenever the cattle have eaten their turnips the byre should be completely cleared of the dung and dirty litter with the graip, shovel, besom and barrow belonging to the byre. A fresh foddering and a fresh littering being given, they should be left to themselves to rest and chew the cud, until the next time of feeding. After eating a little fodder the cattle lie down and rest until visited at night.

Graip or fork

Wheelbarrow

The most personally laborious part of the duty of a cattleman in winter is *carrying straw in large bundles on his back to every part of the steading*. A convenient means of carrying it is with a soft rope about the thickness of a finger, and 3 yards [2.7m] in length, furnished at one end with an iron ring through which the other end slips easily along until it is tight enough to retain the bundle, when a simple loop knot keeps good what it has got. Provided with three or four such ropes, he can bundle the straw at his leisure in the barn, and have them ready to lift when required. The iron ring permits the rope to free itself readily from the straw when the bundle is loosened.

The *dress* of a cattleman is worth attending to, in regard to its appropriateness for his business. Having so much straw to carry on his back, a bonnet or low-crowned hat is most convenient for him; but what is of more importance, when he has charge of a bull, is to have the colour of his clothes of a sombre hue, free of all gaudy or strongly contrasted colours, especially *red*, because that colour from some cause is peculiarly offensive to bulls.

Regularity in regard to time is the chief secret in the successful treatment of cattle. Cattle, dumb creatures though they be, soon understand your plans in regard to what affects themselves, and there is none with which they reconcile themselves more quickly than regularity in the time of feeding; and none on the violation of which they will more readily show their discontent.

Driving and Slaughtering Cattle

Cow droving

It is requisite that cattle which have been disposed of to the dealer or butcher, or are intended to be driven to market, should undergo a preparation for the journey. If they were immediately put to the road to travel, from feeding on grass or turnips, when their bowels are full of undigested vegetable matter, a scouring [diarrhoea] might ensue, which would render them unfit to pursue their journey. This complaint is the more likely to be brought on from the strong propensity which cattle have to take violent exercise on feeling themselves at liberty from a long confinement. They in fact become *light-headed* whenever they leave the hammel or byre, so much so that they actually 'frisk, dance and leap,' and their antics would be highly amusing, were it not for the apprehension they may hurt themselves against some opposing object, as they seem to regard nothing before them.

On being let out for the first time cattle should be put awhile into a large court, or on a road well fenced with enclosures, and guarded by men, to romp about. Two or three times of such liberty will make them quiet; and, in the meantime, to lighten their weight of carcass, they should get hay for a large proportion of their food. These precautions are absolutely necessary for cattle confined in byres, otherwise accidents may befall them on the road, where they will at once break loose. In driving cattle the drover should have *no dog*, which will only annoy them. He should walk either before or behind, as he sees them disposed to proceed too fast or loiter on the road; and in passing carriages the leading ox, after a little experience, will make way for the rest. In other respects their management on the road is much the same as that of sheep, though the rate of travelling is quicker.

Whatever time a lot of cattle may take to go to a market, they should never be *overdriven*. There is great difference in management in this respect among drovers. Some like to proceed on the road quietly, slowly, but surely, and to enter the market in a placid, cool state. Others, again, drive smartly along for some distance and rest to cool awhile, when the beasts will probably get chilled and have a staring coat when they enter the market; while others like to enter the market with their beasts in an excited state, imagining them then to look gay; but distended nostrils, loose bowels and reeking bodies, the ordinary consequences of excitement, are no recommendations to a purchaser.

Assessing a 'ripe' cow

When you look at the *near* side of a *ripe* ox in profile – and this is the side usually chosen to begin with – whatever be its *size*, imagine its body to be embraced within a rectangled parallelogram (see below) and if the ox is filled up in all points, his carcass will occupy the parallelogram;

Hereford bull

but, in most cases, there will be deficiencies in various parts – not that *all* the deficiencies will occur in the same animal, but different ones in different animals.

A greatly commendatory point of a fat ox is a level, broad back from rump to shoulder, because all the flesh seen from this position is of the most valuable description. All that I have endeavoured to describe, in these paragraphs, of the points of a fat ox, can be judged of alone by the eye, and most judges never think of employing any other means; but the assistance derived from the hand is important, and in a beginner cannot be dispensed with.

The first point usually *handled* is the end of the rump at the tail head, although any fat here is very obvious, and sometimes attains to an enormous size, amounting even to deformity. The hook bone gets a touch, and when well covered, is right; but should the bone be easily distinguished, the rump and the loin may be suspected, and, on handling these places, the probability is that they will both be hard and deficient of flesh. To the hand, or rather to the points of the fingers of the right hand, when laid upon the ribs, the flesh should feel soft and thick and the form be round when all is right, but if the ribs are flat the flesh will feel hard and thin, from want of fat. The skin, too, on a rounded rib, will feel soft and mobile, the hair deep and mossy, both indicative of a kindly disposition to lay on flesh. The hand then grasps the flank, and finds it thick, when the existence of internal tallow is indicated. The cod is also fat and large and, on looking at it from behind, seems to force the hind legs more asunder than they would naturally be. The palm of the hand laid along the line of the back will point out any objectionable hard piece on it, but if all is soft and pleasant, then the shoulder top is good. Hollowness behind the shoulder is a very common occurrence; but when it is filled up with a layer of fat, the flesh of all the fore quarter is thereby rendered very much more valuable. You would scarcely believe that such a difference could exist in the flesh between a lean and a fat shoulder. A high narrow shoulder is frequently attended with a ridged backbone, and low-set narrow hooks, a form which gets the appropriate name of *razorback*, with which will always be found a deficiency of flesh in all the upper part of the animal,

A fat long-horned ox (bred and fattened at Dickley Farm, Leicestershire, 1802)

where the best flesh always is. If the shoulder point is covered, and feels soft like the point of the hook bone, it is good, and indicates a well-filled neck vein, which runs from that point to the side of the head. The shoulder point, however, is often bare and prominent. When the neck vein is so firmly filled up as not to permit the points of the fingers into the inside of the shoulder point, this indicates a well-tallowed animal; as also does the filling up between the brisket and inside of the forelegs, as well as a full, projecting, well-covered brisket in front. These are all the *points* that require *touching when the hand is used*; and in a high conditioned ox, they may be gone over very rapidly.

BY-PRODUCTS OF CATTLE

The ox hung up for butchering

Beef is the staple animal food of this country, and it is used in various states: salted, smoked, roasted and boiled. The usual mode of preserving beef is by *salting*; and when intended to keep a long time, such as for the use of shipping, it is always salted with brine: but for family use it should be salted dry with good Liverpool salt, without saltpetre; for brine dispels the juice of the meat, and saltpetre only serves to make the meat dry, and give it a disagreeable and unnatural red colour.

Cattle are useful to man in various other ways than affording food from their flesh – their offal of tallow, hides and horns forming extensive articles of commerce. Hides, when deprived of their hair, are converted into leather by infusion of the astringent property of bark. The old plan of tanning used to occupy a long time; but such was the value of the process that the old tanners used to pride themselves in producing a substantial article. Leather is applied to many important purposes, being made into harness for agricultural and other uses. It is used to line the powder magazines of ships of war, to make carding machines for cotton and other mills; to make belts to drive machinery; to make soles of shoes; and, when japanned, to cover carriages. Calves' leather is used in bookbinding.

'The principal substances of which *glue* is made,' says Dr. Ure, 'are the paring of ox and other thick hides, which form the strongest article; the refuse of the leather dresser; both afford from 45 to 55 per cent, of glue. The tendons, and many other offals of slaughterhouses, also afford materials, though of an inferior quality, for the purpose. The refuse of tanneries – such as the ears of oxen, calves and sheep – are better articles: but parings of parchment, old gloves and, in fact, animal skins in any form, uncombined with tanning, may be made into glue.'

Ox tallow is of great importance in the arts. Candles and soap are made of it, and it enters largely into the dressing of leather and the use of machinery. The horns of oxen and sheep are used for many purposes. The horn consists of two parts: an outward horny case, and an inward conical-shaped substance, somewhat intermediate between indurate [hardened] hair and bone, called the *flint* of the horn. These two parts are separated by means of a blow on a block of wood. The horny exterior is then cut into three portions by means of a frame saw. The lowest of these, next the root of the horn, after undergoing several processes by which it is rendered flat, is made into combs. The middle of the horn, after being flattened by heat and its transparency improved by oil, is split into thin layers, and forms a substitute for glass in lanterns of the commonest kind. The tip of the horn is used by the makers of knife handles and of the tops of whips, and for other similar purposes. The interior, or core, of the horn is boiled down in water. A large quantity of fat rises to the surface; this is put aside and sold to the makers of yellow soap. The liquid itself is used as a kind of glue, and is purchased by the cloth dresser for stiffening. The bony substance which remains behind is then sent to the mill and, being ground down, is finally sold to the farmers for manure.

TREATMENT OF FARM HORSES IN WINTER

With the exception of a few weeks in summer, working farm horses occupy the stall in their stable all the year round (illustrated left). Farm horses are under the immediate charge of the ploughmen, one of whom works a pair and keeps possession of them generally during the whole period of his engagement. This is a favourable arrangement for the horses, as they work much more steadily under the guidance of the same driver, than when changed into unfamiliar hands; and it is also better for the ploughman himself, as he will perform his work much more satisfactorily to himself, as well as his employer, with horses familiarized to him than strange ones.

Stall with cast-iron hind posts

The treatment which farm horses usually receive in winter is this. The ploughmen, when single, get up and breakfast before daybreak, and by that time go to the stable, where the first thing they do is to take out the horses to the water. While the horses are out of the stable the rest of the men take the opportunity of cleansing away the dung and soiled litter made during the night, into the adjoining courtyard with their shovels, wheelbarrow and besoms (shown right). While the horses are absent, usually one of the ploughmen supplies each corn box with corn from the corn chest. It is not an unusual practice to put the harness on while the horses are engaged with the corn, but this should by no means be allowed. Let the horses enjoy their food in peace, as many of them, from sanguine temperament or greed, cannot divest themselves of the feeling that they are about to be taken away from their corn, if worked about during the time of feeding. The harness can be quickly enough put on after the feed is eaten, as well as the horses curried [groomed] and brushed and the mane and tail combed. A very common practice, however, is to dress the horses while eating, which should not be allowed. A better plan in all respects is to let the horses eat their corn undisturbed, and then dress and harness them afterward, and it has the advantage of allowing them a little time between eating their corn and going out to work, which, if of a violent nature, undertaken with a full stomach, may bring on an attack of *baits* or colic.

Square-mouthed shovel

Men and horses continue at work until 12 noon, when they come home, the horses to get a drink of water and a feed of corn and the men, their dinner. Some keep the harness on the horses during this short interval, but it should be taken off to allow both horses and harness to cool and, at any rate, the horses will be much more comfortable without it, and it can be taken off and put on again in a few seconds; and, besides, the oftener the men are exercised in this way they will become the more expert. When the work is in a distant field, rather than come home between yokings, it is the practice of some farmers to feed the horses in the field out of the nose bags; and the men to take their dinners with them, or be earned to them in the field by their own people. This plan may do for a day or two in good weather on a particular occasion; but it is by no means a good one for the horses, for no mode so effectual for giving them a chill could be contrived than to cause them to stand on a head ridge for nearly an hour on a winter day, after working some hours. A smart walk home can do them no harm, and if time is pressing for the work to be done, let the horses remain a shorter time in the stable. The men themselves will feel infinitely more comfortable to get dinner at home.

After dinner they proceed to the stable, when the horses will be found to have finished their feed, and when a small quantity of fodder may be thrown before them fresh from the straw barn, for at this time of year farm horses get no hay. The men may have a few minutes to

Birch or broom besom

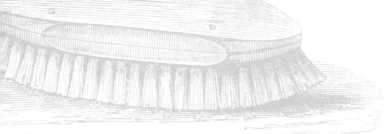

converse until ten minutes to 1pm, when they should give the horses a slight wisp down, put on the harness, comb out their tails and manes and be all ready to put on the bridles the moment 1pm arrives, which is announced by the steward.

The afternoon yoking is short, not lasting longer than sunset, which at this season is before 4pm, when the horses are loosened out of yoke and brought home. After drinking again at the pond they are gently passed through it to wash off any mud from their legs and feet, which they can hardly escape collecting in winter. But in washing the men should be prohibited from wetting their horses above the knees, which they are most ready to do should there be any mud upon the thighs and belly; and to render this prohibition effectual, I have expressly stated, when speaking of the construction of a horse pond, that it should not be made deeper at any part than will take a horse to the knee. There is danger of contracting inflammation of bowels or colic in washing the bellies of horses in winter; and to treat mares in foal – which they will be at this time of year – in this way is little short of madness. If the feet and shanks are cleared of mud, that is all that is required in the way of washing in winter. On the horses entering the stable and having their harness taken off, they are well strapped down by the men with a wisp of straw. After the horses are rubbed down, the men proceed to the straw barn and bundle each four windlings of fodder straw, one to be given to each horse just now, and the other two to be put above the stalls across the small fillets which run along the stable for the purpose.

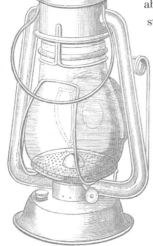

Stable lantern

When 8pm arrives, the steward, provided with light in a lantern, summons the men to the stable to give the horses a grooming for the night and their suppers. The sound of a horn, or ringing of a bell, is the usual call on the occasion, which the men are ready to obey. Lights are placed at convenient distances in the stable to let the men see to groom the horses. The grooming consists first in currying the horse with the curry comb, to free him of all dirt that may have adhered to the skin during the day, and which has now become dry and flies off. A wisping of straw removes the roughest of the dirt loosened by the curry comb (shown below). The legs ought to be thoroughly wisped, not only to make them clean, but dry of any moisture that may have been left in the evening, and at this time the feet should be picked clear by the foot picker of any dirt adhering around between the shoe and the foot. The brush is then used to remove the remaining and finer portions of dust, from which, in its turn, it is cleared by a few rasps of the curry comb. The straw of the bedding is then shaken up with a fork, the horses then get their feed of oats, after which the lights are removed and the stable doors barred and locked by the steward, who is custodian of the key.

This is the usual routine of the treatment of farm horses in winter and, when followed with some discernment in regard to the state of the weather, is capable of keeping them in health and condition. The horses are themselves the better of being out every day, but the species of work which they should do daily must be determined by the state of the weather and the soil. In very wet, frosty or snowy weather, the soil cannot be touched; but then thrashing and carrying corn to market may be conducted to advantage, and the dung from the courts may be taken out to the fields in which it is proposed to make dung hills. This latter piece of work is best done when the ground is frozen hard. When heavy snow falls, nothing can be done out of doors with horses, except threshing when the machine is impelled with horse power. On a very rainy day, the horses should not go out, as everything about them, as well as the men, become soaked; and before both or either can be again made comfortable, the germs of serious disease may be laid in both. When it is fair above, on the other hand, however cold the air or wet the soil, some of the sorts of outdoor work mentioned above may be done by the horses; and it is better for them to work only one yoking a day than to stand idle in the stable.

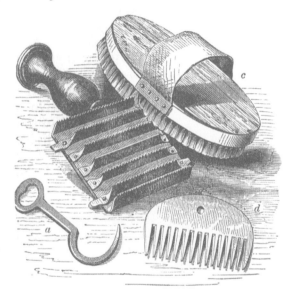

Foot picker (*a*), curry comb (*b*), brush (*c*) and mane comb (*d*)

Fattening, Driving and Slaughtering Swine

Fattening pigs

A smock-clad pig farmer and his daughter with his pigs at feeding time

There should be no littering of young pigs in winter on an ordinary farm, because all young pigs are very susceptible to cold; and as chances of being exposed to it will frequently happen in the most comfortable sty, it is scarcely possible to avert its injuries; which are, reddening of the skin, causing the coat to stare, and if not actually killing the pigs,

chilling them to such a degree as to prevent their growth until the
return of more genial temperature in spring. If circumstances
however, render it profitable to raise suckling pigs at
Christmas – and a roast pig at that joyous season is a
favourite dish in England – the matter may be
accomplished as easily as the raising of house lamb at
that season, by having sties for the sows under a closed
roof, with doors and windows to shut out cold and admit light.

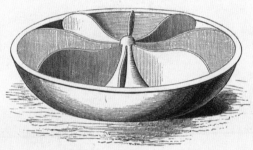

Common straw-fork

That the youngest pigs may receive better treatment, the court and
shed are made purposely for them. These pigs consist, probably, of the
last litters of the season of as many brood sows as
are kept. Here they should be provided daily
with turnips as their staple food, of the
sort given for the time to the cattle, and
sliced as small as for sheep; and they
should, besides, have a portion of a
warm mash made, with such other
pickings from the farmhouse which the
kitchen affords. They should also be
provided with a trough of clean water, and
plenty of litter under the shed every day. The
courtyard should be cleaned out every day. Pigs are accused of dirty
habits, but the fact is otherwise, and the accusation applies more truly
to their owners who keep them dirty, than to the natural habits of the
animals themselves.

Ring pigs' trough, to stand in court

It is the duty of the cattleman to supply the store pigs with food and
clean out their courtyard, and this part of his duty should be conducted
with as much regularity as the feeding of cattle. Whatever food or drink
may be obtained from the farmhouse for the pigs is brought to the court
by the dairymaid.

The older pigs have the liberty of the large courts where they make
their litter amongst the cattle in the open court, when the weather is
mild, and in the shed, when it is cold. Though thus left at liberty, they
should not be neglected of food, as is too often the case. They should
have sliced turnips given them every day, in troughs, and they should
also have troughs of water. Pigs, when not supplied with a sufficiency of
food, will jump into the troughs of the cattle, and help themselves to

turnips; but this dirtying of the cattle's food and troughs, they should be made to understand by coercion, is a practice not to be tolerated. The cattle man attends upon these pigs, giving them sufficiency of turnips and water. An excellent form of trough is the ring pigs' trough, fitted to stand in the middle of a large court, and to contain the food of a number of pigs (see the illustration on page 83). It stands upon the dunghill, is not easily overturned, the cattle cannot hurt themselves upon it, and it keeps the food subdivided into a number of compartments; while it can easily be pushed about to the most convenient part of the court for supplying the food.

The farmer should once a year fatten a few pigs for his own use as ham. These should be *at least a year old*, attain the weight of 18 or 20 stones [114–127kg], and be slaughtered about Christmas. Castrated males or spayed females are in the best state for this purpose. Four pigs of 20 stones [127kg] each every year will supply a pretty good allowance of ham to a farmer's family. Up to the time of being placed in these sties, the pigs have been treated as directed above; but when confined, and intended to be *fattened to ripeness*, they receive the most nourishing food.

ASSESSING A FAT PIG

Of *judging* of a fat pig, the back should be nearly straight; and though arched a *little* from head to tail, it is no fault. The back should be uniformly broad and rounded across along the whole body. The touch all along the back should be firm but springy, the thinnest skin springing most. The shoulder, side and hams should be deep up and down, and in a straight line from shoulder to ham. The closing behind should be filled up; the legs short and bone small; the neck short, thick and deep; the cheeks rounded and well filled out; the face straight, nose fine, eyes bright, ears pricked, and the head small in proportion to the body. A curled tail is a favourite, because indicative of a strong back. A *fat pig* ought never to be driven, but carried in a cart when desired to be transported from one place to another.

By-products of pigs

Hog's lard is rendered in exactly the same manner as mutton suet; but as lard is liable to become rancid, yellow-coloured, and acquire a strong smell when exposed to the air, it is usually tied up in bladders. For this purpose, it is allowed to cool a while, after it is melted, and the bladder (a pig's or calf's) being made ready by being thoroughly cleaned and turned outside in, is filled with the lard by a funnel and tied up.

Hog's skin is usually thick and, when tanned, its great toughness renders it valuable for the seats of riding saddles. Hog's bristles are formed into brushes for painters and artists, and for numerous domestic uses. Some of the offals of the pig make excellent domestic dishes, such as blood, mealy and sweet puddings; and pork sausages, made of the tender muscle under the lumbar vertebrae, are sweeter, higher flavoured and more delicious than those of beef.

Berkshire boar

THE TREATMENT OF FOWL IN WINTER

A trusted dog keeps guard in the poultry yard

Of all the animals reared on a farm, there are none so much neglected by the farmer, both in regard to the selection of their kind, and their qualifications to fatten, as all the sorts of domesticated fowls found in the farmyard. The usual objection urged against feeding fowls is that it does not pay, and they are thus considered worthless stock. But it is as much the mode of managing them that renders them so. Apart from every consideration of profit to be derived from sales in market towns, there is the superior one of the farmer having it at *all times* in his power

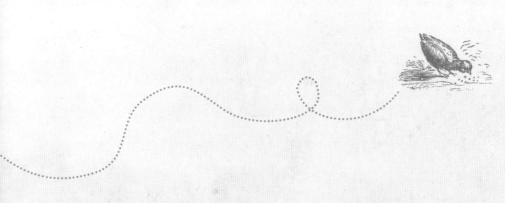

to eat a well-fed fowl at his own table; and there is no good reason why he should not be able to enjoy such a luxury at any time he chooses.

The ordinary fowls on a farm are the cock *(Phasianus gallus)*, the turkey *(Meleagris gallopavo)*, the goose *(Anas anser)*, the duck *(Anas domesticus)*, the pigeon *(Columba livia)* and the white-backed or rock dove, which was long confounded with the blue-backed dove *(Columba oenas)*.

Hens

And first, in regard to the condition of the *hen*. As hatchings of chickens are brought out from April to September, there will be broods of chickens of different ages in winter; some as old as to be capable of laying their first eggs, and others only mere chickens. The portion of those broods which should be taken for domestic use are the young cocks and the older hens, there being a feeling of reluctance to kill young hens, which will supply eggs largely in the following season. At all events, should any hen chickens be used for the table, the most likely to become good layers next season should be preserved. The marks of a chicken likely to become a good hen are a small head, bright eyes, tapering neck, full breast, straight back, plump ovoid-shaped body and moderate-length grey-coloured legs. All the yellow-legged chickens should be used, whether male or female, as their flesh never has so fine an appearance as the others. Young fowls may either be roasted or boiled, the male making the best roast of all, and the female the neatest boil. The older birds may be boiled by themselves, and eaten with bacon, or assist in making broth, or that once favourite winter soup in Scotland, *cockieleekie*.

The criterion of a *fat* hen is a plump breast, and the rump feeling thick, fat and firm, on being handled laterally between the finger and thumb. A corroborative criterion is thickness and fatness of the skin of the abdomen, and the existence of fat under the wings. White flesh is always preferable, though poulterers insist that a *yellow-skinned* chicken makes the most delicate *roast*. A hen is deprived of life by dislocation of the neck by drawing, where the blood collects and coagulates.

Turkey

Turkeys being hatched in May, will be full grown in stature by winter and, if they have been well fed in the interval, will be ready for use. Indeed, the Christmas season never fails to create a large demand for turkeys, and it must be owned there are few more delicate and beautiful dishes presented at table, or a more acceptable present given to a friend, than a fine turkey. Young cocks are selected for roasting, and young hens for boiling, and both are most relished with a slice of ham, or of pickled ox tongue.

The criterion of a good turkey is fullness of the muscles covering the breast bone, thickness of the rump and existence of fat under the wings; but the turkey does not yield much fat, its greatest property being plenty of white flesh. A turkey is deprived of life by cutting its throat, when it becomes completely bled.

Geese

Geese, having been hatched in the early part of summer, will also be full grown and fit for use in winter. The criterion of a fat goose is plumpness of muscle over the breast, and thickness of rump when alive; and, when dead and plucked, the additional one of a uniform covering of *white* fat over the whole breast.

Geese are always roasted; and their flesh is much heightened in flavour by a seasoning of onions as a stuffing, and by being served up

with apple sauce. A goose should be kept a few days before being used. It is bled to death by an incision across the back of the head, which completely bleeds it.

Geese have long been proverbially good watchers. I have seen a gander announce the approach of beggars towards the kitchen door as lustily as any watchdog.

Ducks

Ducks, being also early hatched, are in fine condition in winter, if they have been properly fed. Ducklings soon become fit for use, and are much relished with green peas in summer.

They are most frequently roasted, and stuffed with sage and onions; though often stewed and, if smothered among onions when stewed, there are few more savoury dishes that can be presented at a farmer's table. A duck never eats better than when killed immediately before being dressed. It is deprived of life by chopping off the head with a cleaver, which completely bleeds it.

Apples are picked in preparation for the stuffing of the traditional Michaelmas goose

Housing and feeding poultry

The hen houses are divided into three apartments, each having a giblet check door to open outwards, and all included within a courtyard, provided with an outer door and lock. The use of three apartments is to devote one of them to the hens and turkeys, which roost high, another to the geese

'The utility of domestic fowls to man in their employment during life and uses after death'. 1845

and ducks, which rest on the floor, and the third to a hatching house to accommodate both. When geese are obliged to rest below hens, they are made uncomfortable and dirty by the droppings of those which roost above them.

The daily treatment of fowls may be conducted in this manner – Some person should have special charge of them, and the dairymaid is perhaps the best qualified for it. As fowls are very early risers, she should go to the hen house in the morning, on her way to the byre, and let out all the fowls, giving the hens and turkeys a feed of light corn and cold boiled potatoes, strewed along at some convenient and established place out of the way of the general passage of horses and carts.

The ducks should get the same food either near the horse pond, or where there is a pond or trough of water, as they cannot swallow dry food without the assistance of water. Geese thrive well upon sliced turnips, a little of which, sliced small, should be left by the cattle man for the dairymaid at any of the stores, and given at a place apart from the hens. When stated places are thus established for feeding fowls at fixed hours, they will resort to them at those hours; at least the well-known call will bring the hour to their recollection, and collect them together on the spot in a few seconds, and the regular administration of food being as essential for their welfare as that of other stock.

After her own dinner, say 1pm, the dairymaid takes a part of the potatoes that have been boiled at that time, and while a little warm, gives them crumbled down, from their skins, with some light corn, to the turkeys and hens.

Before sunset, the fowls are all collected together by a call, and put into the house, and which they will readily enter; and many will have taken up their abode in it already, especially the turkeys, which go very soon to roost. The ducks are the latest idlers. The floors of the different apartments should be littered with a little fresh straw every day, sufficient to cover the dung, and the whole cleaned out every week. Sawdust or sand, where they are easily obtained, form excellent covering for the floor of hen houses. Troughs of water should be placed in the courtyard, and supplied fresh and clean every day.

Poultry house on wheels

Eggs

I have said that *eggs*, and chickens too, may be obtained in winter by good management. The young hens of the first broods in April will be old enough to lay eggs in winter. A few of these should be selected for the purpose; and when the period of laying approaches – which may be ascertained by their chaunting a song and an increased redness of the comb – they should be encouraged by better feeding and warmer housing at night. These three or four young hens will lay as many eggs every day; and though they are not so large as those of more matured fowls, being only pullets' eggs, still they will be fresh; and it is no small luxury to enjoy a new-laid egg at breakfast every winter morning.

THRASHING AND WINNOWING GRAIN, AND THE THRASHING MACHINE

Ladder, 15 feet [4.6m] long

The first preparation made for *thrashing* corn – that is, separating the grain from the straw by the thrashing machine or the flail – is casting in the stack to be thrashed, and mowing it in the upper or thrashing barn. When about to *cast* a stack, he provides himself with a ladder to reach its eaves, and a long small fork usually employed to pitch sheaves at leading time to the builder of stacks. He also provides himself with a stout clasp knife, which most farm servants carry. Standing on the ladder, he, in the first place, cuts away with the knife all the tyings of the straw ropes at the eaves of the stack. On gaining the top, the ladder is taken away, and he cuts away as much of the ropes as he thinks will allow him to remove the covering with the fork. The covering is then pushed down to the ground, until the top of the stack is completely bared. On the side of the stack nearest the barn, a little of the covering is spread upon the ground by the fieldworkers, to keep the barn sheet off the ground, and they spread it over the spread straw, close to the bottom of the stack. This *sheet* consists of thin canvas, about 12 feet [3.7m] square. The sheaves first thrown down from the top of the stack upon the sheet are taken by the women, and placed side by side, with the corn end upon the sheet, along both its sides, to keep them down from being blown up by the wind, or turned up by the feet.

A fourth worker remains in the upper barn, to pile up the sheaves as they are brought in into what are called *mows*, that is, the sheaves are placed in rows, parallel to each other to a considerable height, with their butt ends outwards, the first row being piled against the wall. In casting the stack, the steward takes up the sheaves in the reverse order in which the builder had laid them at harvest time, beginning with those in the centre first, and then removing those around the circumference one by one. The fork thrust into the band will generally hit the centre of gravity of the sheaves, where they are most easily lifted, and swung towards the sheet. When all the sheaves of the stack have been wheeled in, the steward takes a rake and clears the ground of all loose straws of corn that may have become scattered around the base of the stack, and puts them into the sheet, the four corners of which are then doubled in towards the middle, including within them the grain that may have been shaken out by the shock received by the sheaves on being thrown down; and the sheet, with its contents, are carried by all the women into the barn, and its contents emptied on the floor, near the feeding in board. The sheet is then shaken, and spread out upon the stackyard dyke, or other airy place, to dry before being folded up to be ready for use on a

Wooden rick stand

Feeding sheaves into the thrashing machine

Corn barrow

Corn basket of wickerwork

similar occasion. The covering of the stack is then carried away by the women, to such parts of the courts and hammels as are considered by the cattleman to require littering, before it becomes wetted with rain, and the ground raked clean. The straw ropes, which bound down the covering of the stack, should be cut by the steward *into short lengths* before being carried away in the litter, as *long* ropes are found very troublesome to the men when filling their carts with dung on clearing out the courts.

Everything being thus prepared (and every preparation ought to be completed before the mill is moved), the mill is ordered by the steward to be set a-going by the engine man or driver, when the power is steam or horses, and he himself lets on the water to the wheel when the power is water. The power should be applied gently at first, and no corn should be presented until the mill has acquired its proper momentum, the *thrashing motion*, as it is termed. When this has been attained, which it will be in a very few seconds, and which a little experience will teach the ear to recognise instantly, the steward – the feeder-in – takes a portion of a sheaf in both his hands, and letting its corn end fall before him on the feeding-in board, spreads it with a shaking and disengaging motion across the width of the board. His great care is that no more is fed in than the mill can thrash cleverly; that none of the corn is presented sideways, or with the straw end foremost. He thus proceeds with a small quantity of corn for a few minutes, until he ascertains the capacity of the mill for work at the particular time, when the quantity required is fed in; but this, on any account, should never exceed one sheaf at a time, however fast they may have to be supplied in succession afterwards.

Proceeding in this way, the feeder in *d* (page 93), takes the sheaves from the table *e*, which are supplied him by a woman stationed beside it, whose duty it is to loosen the bands of the sheaves; but he should not allow her to put on more than one sheaf at a time on the table, as is the

Portable threshing machine, showing straw elevator

propensity to do, much to his annoyance; while the other woman *g*, brings forward the sheaves and places them in a convenient position before the other woman *f*, and even loosens one occasionally in her assistance.

The state of the corn in reference to dryness or dampness, and lengthiness and shortness of the straw, as well as incidents in the moving power, will very much affect the progress of thrashing. When the sheaves are long, the feeding rollers will of course take a longer time to take them in; and to make them take in faster, the fast motion should be given to them. A slower motion than requisite is apt to chop the straw in pieces by the scutchers of the drum. On the other hand, short sheaves may be taken in so quickly as not to afford the drum the requisite time to separate the corn from the straw; in which case the rollers should be put upon the slow motion.

The first thing to be done towards preparing the thrashed heap of corn for the market is passing the roughs through the blower, or winnowing machine, or fanners. This machine is set with its tail at the barn door, that the chaff blown away may fall upon the causeway of the

Barn stool

court, and be kept out of the corn barn. The steward drives the fanners, one woman fills the hopper with the roughs, either of wheat or oats, for the barley roughs, as you have seen, have been put through the mill again; and as roughs do not pass easily through the hopper, another woman stands upon the stool belonging to the barn, and feeds them in with her hand towards the shoe or feeding roller, while the other two women riddle the corn upon the heap that had been riddled from the clean spout of the thrashing mill. The riddlings of the roughs, and all the light corn, may be put into an enclosed space, such as at the bottom of the granary stair in the corn barn to be fed to the fowls.

This matter being disposed of, the heap of grain, suppose it to be wheat, is next to be winnowed (see below). For this purpose, the blower is placed alongside the heap, with its tail away from the direction in which it is proposed to place the new-riddled heap of grain with its offside, that is, its side farthest from the driver, next to the heap. The steward adjusts the component parts of the blower to suit the nature of the grain to be winnowed, namely, the tail board should be no higher up than to allow the chaff to escape over it. While it retains the lightest

Winnowing corn

even of the grain, the slide in the interior should only be so far up as to permit the light grain to be blown over it, while it retains all the heaviest, which pour down onwards to the floor. What falls from this slide is the light corn, and it drops nearest the chaff. The wire screen below this slide permits dust and small seeds of wild plants to pass through, and deposits them between the light and heavy corn. The opening at the feeding roller is so adjusted as that the grain shall fall as fast, but no faster than the wind shall have power to blow away the chaff and light corn from amongst the heavy. All these adjustments of parts may not be made the most perfect at once, but a little trial will soon direct him what requires to be rectified, and experience of the machine will enable him to hit near the mark at once. The blower should be made to stand firmly and steadily on the floor when used.

The arrangement of persons for winnowing corn, so as to proceed with regularity and dispatch, is this. The steward *(b)* drives the blower. One woman fills the hopper with corn with a large basket from the heap *(c)*, on the opposite side from the driver. Her duty is to keep the hopper as nearly full as she can, as then the issue of corn from it is most regular. Another woman *(d)*, with a smaller basket, takes up the good grain as it slides down at the end, and divides the basketful between the other two women *(e)* who stand with a riddle each in her hand at the place where the new heap is to be made. The heap is made in one corner *(g)*, or against any part of a wall of the barn, to take up as little room as possible. When the two women have received the grain into their riddles, they riddle it, bringing the last part of each riddling towards the edge of the heap, and casting what is left as the scum in the riddles into the bushel *(f)* placed conveniently to receive it. The riddlings consist of capes, large grains, sprouted grains, small stones, the larger class of seed of weeds that could not pass through the wire screen in the blower, clods of earth, bits of straw too heavy to be blown away and such like. By the time the women have riddled the quantity given them, the other woman is ready to

Sack barrow

supply them with a fresh quantity. When the corn begins to accumulate amongst the riddlers' feet, one of them takes the wooden scoop (shown below) and, drawing with it the tail or edge of the heap into a small heap, gives it up in portions to the other riddler, who puts the remains of the riddlings into the bushel; after which the large heap is shovelled up against the wall, while the scattered grain on the floor is swept towards it with a besom, by the other riddler, or the woman who gives up the corn from the blower, as the case may be. While the unwinnowed heap is becoming less, as the riddled one increases in bulk, the woman who has charge of it shovels it up at times, and sweeps in the edge, that no scattered grains may be permitted to lie upon the floor to get crushed with her shoes. All the women should endeavour to do their respective parts in a neat and cleanly way. There is much difference in the mode of working evinced by different women in the barn, some constantly spilling grain on the floor, when they have occasion to lift it with a wecht, evincing the slattern; but it is the duty of the steward to correct every instance of carelessness; whilst others keep the floor clean, and handle all the instruments they use with commendable skill and neatness.

The thrashed heap of corn being thus passed through the blower, and riddled in the manner described into another heap, the chaffy matter blown upon the floor is then carried away to the dunghill, and the light corn subjected to examination, as well as the riddlings in the bushel. When the grain is of fine quality, there will be no good grain, and little bulk in the light corn heap, which may all be put past for hen's meat; but in other circumstances the light corn, together with what is in the bushel, should again be put through the fanners, and the grain taken out of it that would not injure the clean corn, if mixed with it. When the light corn has thus been disposed of, and the seeds and dust from the screen carried out and placed on a bare piece of ground for the pigeons, fowls, or wild birds to pick up, and *not* thrown upon the dunghill to render it foul with the seeds of wild plants, the heap should be shovelled up, the fanners thoroughly made clean and placed aside, and the floor swept.

Wooden corn scoop

Wheat

On examining the ears of *wheat* that have come under my notice, I think they may be divided into the three classes (see the illustration) and which may be distinguished thus: *a* is a close or *compact-eared* wheat, which is occasioned by the spikelets being set near each other on the rachis or jointed stem, and this their position has a tendency to make the *chaff short and broad*, and the spikelets are so also. This specimen of the close-eared wheat is Hickling's Prolific. The second class of ears is seen at *b*, the spikelets being of *medium* length and breadth, and placed just so close upon the rachis as to screen it from view. The ear is not so broad, but longer than *a*. The *chaff* is of *medium length and breadth*. This specimen is the well-known Hunter's white wheat. The third class is seen at *c*, the spikelets of which are set *open*, or so far asunder as to permit the rachis to be easily seen between them. The ear is about the same length as the last specimen, but is much narrower. The *chaff* is *long and narrow*. This is a specimen of Le Couteur's Bellevue Talavera white wheat.

Classification of wheat by the ear

In *d* is represented a bearded wheat, to show the difference of appearance which the beard gives to the ear. The bearded wheats are generally distinguished by the *long shape of the chaff* and the open position of the spikelets, and therefore fall under the third class. But cultivation has not only the effect of decreasing the strength of the beard, but of setting the spikelets closer together, as in the specimen of the white Tuscany wheat, shown at *d* in the cut, which is considered the most compact-eared and improved variety of bearded wheat. Bearded wheat constitutes the second division of cultivated wheat of the botanists, under the title of *Triticum sativum barbatum*. The term bearded wheat is used synonymously with spring wheat, but erroneously, as some beardless wheat is as fit for sowing in spring as bearded, and some bearded may be sown in winter.

Barley

Classification of barley by the ear

Barley's botanical position is the third class *Triandria*, second order *Digynia*, genus *Hordeum* of the Linnaean system. Professor Low divides the cultivated barleys into two distinctions, namely, the two-rowed and the six-rowed, and each of these comprehends the ordinary, the naked and the sprat or battledoor forms. Mr Lawson describes 20 varieties of barley, while the Museum of the Highland and Agricultural Society contains specimens of 30 varieties. The natural classification of barley by the ear is obviously of three kinds, four-rowed, six-rowed, and two-rowed (shown left). Where the three forms are represented, *a* is the four-rowed, or bere or bigg; *c* is the six-rowed; and *b*, the two-rowed. Of these the bere or bigg was that which was cultivated until a recent period, when the two-rowed has almost entirely supplanted it, and is now the most commonly cultivated variety, the six-rowed being rather an object of curiosity than culture.

Oats

The tartarian oat; the potato oat

Oats are cultivated on a large extent of ground in Scotland, and it is believed that no country produces greater crops of them or of finer quality. The plant belongs to the natural order of *Gramineae*, and it occupies the third class *Triandria*, second order *Digynia*, genus *Avena*, of the Linnaean system. Its ordinary botanical name is *Avena sativa*, or cultivated oat. The term oat is of obscure origin. Paxton conjectures it to have been derived from the Celtic *etan*, to eat. There are a great number of varieties of this grain cultivated in this country. Mr Lawson describes 37; and 54 are deposited in the Highland and Agricultural Society's Museum. The natural classification of this plant by the *ear* is obvious. One kind has its panicles spreading and equal on all sides, and tapering towards the top of the spike in a conical form. The other has its panicles shortened, nearly of equal length, and all on the same side of the rachis. Both kinds are represented here, where *a* shows the first, and *b*, the second, where they both appear somewhat confined or squeezed towards the rachis, the object being to exhibit the grain in the straw as taken from the stack, rather than when pulled green from the field. The prolific potato oat, which is beardless, is commonly cultivated in Scotland for the sake of its meal, while *b* is the white Tartarian oat, which is bearded, and is extensively cultivated in England for horse corn.

RYE

Botanically, *rye* occupies the same place, both in the natural and sexual systems, as the other grains which have been described. It is the *Secale cereale* of the botanists, so called, it is said, *à secando*, from cutting, as opposed to leguminous plants, whose fruits used to be gathered by the hand. It is a narrow small grain, not unlike shelled oats. There is only one known species of this plant, which is said to be native of Candia [now Crete], and was known in Egypt 3,300 years ago; but there are several varieties which are raised as food. Rye is not much cultivated in this country, where only a patch here and there is to be seen. It is, however, extensively cultivated on the Continent, especially in sandy countries.

Ear of rye

BEANS

Beans belong to a very different tribe of plants to those we have been considering. They belong to the natural order of *Leguminosae*, because they bear their fruit in legumes or pods; and in the Linnaean system they occupy the class and order *Diadelphia decandria*. Their generic term is *Faba vulgaris*; formerly they were classed amongst the vetches and called *Vicia Faba*. The common bean is divided into two classes, according to the mode of culture to which they are subjected, that is, the field or the garden. Those cultivated in the field are called *Faba vulgaris arvensis*, or as Loudon calls them, *Faba wlgaris equina*, because they are cultivated chiefly for the use of horses, and are usually termed horse beans. The garden bean we have nothing to do with, though some farmers attempt some of the garden varieties in field culture, but I believe without success.

Horse beans

STRAW

WHEAT STRAW

Wheat straw is generally long. The strength and length of wheat straw render it useful in thatching, whether houses or stacks. It is yet much employed in England for thatching houses, and perhaps the most beautiful thatches are to be found in the county of Devon. Wheat straw makes the best thatching for corn stacks, its length and straightness ensuring safety, neatness and dispatch in the process, in the busy period of securing the fruits of the earth. It forms an admirable bottoming to the bedding in every court and hammel of the steading. As littering straw, wheat straw possesses superior qualities. It is not so suited for fodder to stock, its hardness and length being unfavourable to mastication; yet I have seen farm horses very fond of it. Horses in general are fond of a hard bite, and were wheat straw cut for them by the chaff cutter of a proper length, I have no doubt they would prefer it to every other kind of straw as fodder. The *chaff* of wheat does not seem to be relished by any stock; and is therefore strewn on the dunghill, or upon the lairs of the cattle within the sheds.

BARLEY STRAW

Barley straw is always soft, and has a somewhat clammy feel, and its odour, with its chaff, when new-thrashed, is heavy and malt-like. It is relished by no sort of stock as fodder; on the contrary, it is said to be deleterious to horses, on whom its use is alleged to engender grease in the heels. Barley chaff, however, is much relished by cattle of all ages, and rough as the awns are, they never injure their mouths in mastication. Barley straw is thus only used as litter, but it is much inferior to wheat straw either for cleanliness, durability or comfort.

Oat straw

Oat straw is most commonly used as fodder, being considered too valuable to be administered in litter. It makes a sweet, soft fodder and, when newly thrashed, its odour is always refreshing. Oat straw is very clean, raising little or no dust, and so is its chaff; and on this account, as well as its elasticity, the latter is very commonly used in the country to make beds with tickings, for which purpose the chaff is riddled through an oat riddle, and the larger refuse left in the riddle thrown aside. Sheep are very fond of oat straw, and will prefer it to bad hay.

Rye straw

Rye straw is small, hard and wiry, quite unfit for fodder, and perhaps would make but uncomfortable litter in a stable, though it would, no doubt, be useful in a court for laying a durable bottoming for the dunghill; but it forms most beautiful thatch for houses, and would, of course, do for stacks, if it were not too expensive an article for the purpose. It is much sought for by saddlers for stuffing collars of posting and coach horses, and in default of this wheat straw is substituted.

Pease and bean straw

It is difficult in some seasons to preserve the straw, or *haulm*, of the pulse crops, but, when properly preserved, there is no kind of straw so great a favourite as fodder with every kind of stock. An ox will eat pease straw as greedily as he will hay; and a horse will chump bean straw with more gusto than ill-made rye-grass hay. Young cattle are exceedingly fond of bean and pease chaff. These products of the pulse crops are considered much too valuable to be given as litter. Since bean chaff is so much relished by cattle, there is little doubt that bean and pea haulm, cut into chaff, would be relished, and were this practice attended to in spring, the hay usually given to horses at that season might be dispensed with on farms which grow beans and peas. It is said that when work horses are long kept on bean straw their wind becomes affected.

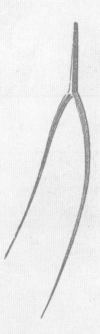

Common straw fork

THE FORMING OF DUNGHILLS

Towards the close of winter, the dung will have accumulated so high in the large courts as to become nearly level with the feeding troughs and thereby making them inconveniently low for the cattle; but before this inconvenience occurs, the dung should be removed and formed into dunghills in the fields intended to be manured in the ensuing season. The court dunghill contains the litter of the work-horse stable, and the pigsty; and the litter of the servants' cow byre, besides its own.

The litter in the court will be found much compressed in consequence of heavy cattle being obliged to move over it frequently within a limited space. It is sometimes so compressed as almost to resist the entrance of the graip (dung fork). To enable it to be easily lifted, it should be cut in parallel portions with an implement called the dung spade (shown right). This consists of a heart-shaped blade of steel thinned to a sharp edge along both sides and its helve [handle] with a cross head is fastened with nails in a split socket. In using this spade it is lifted up with both hands by the cross head and its point thrust with force into the dung heap, and it is then used like a common spade while rutting turf, with the foot upon the upper part of the blade.

There is another matter which deserves consideration before the court is begun to be cleared of its contents; which is the position the dunghill or dunghills should occupy in the field, and this point is determined partly by the form which the surface of the field presents, and partly from the point of access to the field. In considering this point, which is of more importance than it may seem to possess, it should be held as a general rule that the dunghill should be placed where the horses will have the advantage of going downhill with the loads from it.

The fields to which the dung should be carried are those to be fallowed in the ensuing season; that is, set apart for the growth of green crops, such as potatoes and turnips, and for the part which receives more cleaning than a green crop admits of, namely, a bare fallow. The potato culture coming first in order, the land destined for that crop should have its manure carried out and formed into the first dunghill. The turnips next come in hand; and then the bare fallow. The dunghills intended for potatoes and turnips should of course be made respectively of such a size as to manure the extent of land to be occupied by each crop. The manure for bare fallow not being required till much farther on in the season, may be deferred being carried out at present.

On forming a dunghill in the field, some art is requisite. One of a breadth of 15 feet [4.6m], and of four or five times that length, and of proportionate height, will contain as much manure as should be taken from one spot in manuring a field quickly. Suppose that 15 feet [4.6m] is fixed upon for the width, the first carts should lay their loads down at the nearest end of the future dunghill, in a row across the whole width, and these loads should not be spread thin. Thus, load after load is laid down in succession upon the ground maintaining the fixed breadth, and passing over the loads previously laid down. After the *bottom* of the dunghill has thus been formed of the desired breadth and length, the further end is then made up with layer after layer, until a gradual slope is formed from its nearest to its farthest extremity. This is done with a view to effecting two purposes; one, to afford an easy slope for the loaded carts to ascend, the other, to give ease of draught for horses and carts to move along the dunghill in all parts, in order to compress it firmly. Every cartload laid down above the bottom layer is spread around, in order to mix the different kinds of dung together, and to give a uniform texture to the manure. To effect this purpose the better, a field worker should be employed to spread the loads on the dunghill, as they are laid down. When the further end has reached the height the dunghill is thought will contain of the desired quantity of manure, that height is brought forward towards the nearest end; but the centre of the dunghill will necessarily have the greatest elevation, because a slope at both *ends* is required to allow the carts to surmount the dunghill and then to come off it. It is an essential point to have the whole dunghill equally compressed, with the view of making the manure of similar quality throughout. After the carting is over, the scattered portions of dung around the dunghill should be thrown upon the top, and the top itself levelled along and across its surface.

Dung spade

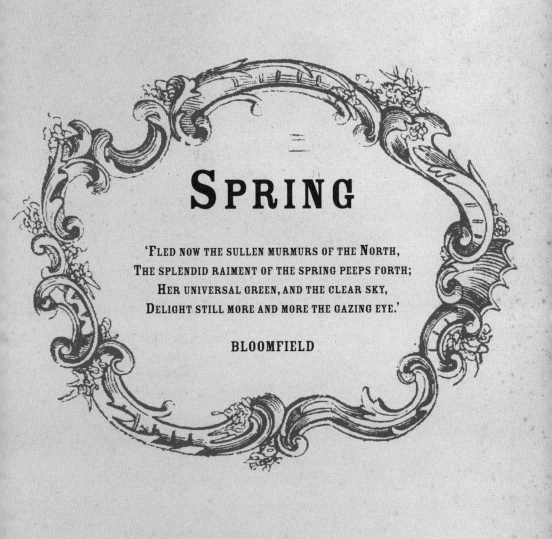

Spring

'Fled now the sullen murmurs of the North,
The splendid raiment of the spring peeps forth;
Her universal green, and the clear sky,
Delight still more and more the gazing eye.'

BLOOMFIELD

NTRODUCTION

Upon the whole, we have seen that winter is the season of repose, of passive existence, of dormancy, though not of death. Spring calls forth the opposite emotions; it is the season of revivification, of passing into active exertion, of hope, nay, of confidence that what we do will succeed – of hope ripening into fruition as the earnest prospective of plenty is presented in the reproduction of the herds and flocks, and in the world of life which springs into view immediately after the industrious hand has scattered the seed upon the ground. The joy in contemplating such a prospect to the issue of labour is indescribable.

The weather in spring, in the zone we inhabit, is exceedingly variable, alternating, at short intervals, from frost to thaw, from rain to snow, from sunshine to cloud. The sky is very clear when the air is free of clouds. The winds are very sharp, when coming from the north or northeast direction; and they are frequent, blowing strongly sometimes from an eastern and sometimes from a western direction. In the former they are piercing, even though not inclining to frost; in the latter they are strong, boisterous, squally and rising at times into tremendous hurricanes, in which trees only escape being uprooted in consequence of their leafless state. The air, when dry, evaporates moisture quickly; and the surface of the ground is as easily dried as wetted. Very frequently snow covers the ground for a time in spring.

Spring is a busy season on the farm. The cattleman, besides continuing his attendance on the fattening cattle, has now the more delicate task of waiting on the cows at calving, and providing comfortable lairs for new-dropped calves. The dairymaid commences her labours, not, it is true, in the peculiar avocations of the dairy, but in rearing calves – the supply of a future herd – which, for a time, are indulged with every drop of milk the cows can yield. The farrows of pigs now claim a share in her solicitude. The shepherd, too, has his painful watchings, day and night, on the lambing ewes; and his solicitude and tenderness for the simple lambs, until they are able to frisk and gambol upon the new grass, is a scene of peculiar interest, and insensibly lead to higher thoughts.

The condition of the fields demands attention as well as the reproduction of the stock. The day now affords as many hours for labour as are usually bestowed at any season in the field. The ploughmen, therefore, know no rest for at least 12 hours every day, from the time the harrows are yoked for the oat seed until the potato and turnip crops are sown. The beans first demand the ploughmen's aid, and then the lea ground, turned over at intervals of fresh weather in winter, is ready, with a due degree of drought, to receive its allowance of oat seed. The turnip land, bared as the turnips are consumed by sheep, is now ploughed across, or ridged up at once for spring wheat, should the weather be mild and the soil dry enough, or else the ridging is delayed for the barley seed. The fields containing the fallow land now receive a cross furrow, in the order of the fallow crops, the potatoes first, then turnips, and lastly the bare fallow. Grass seeds are now sown amongst the young autumnal wheat, as well as amongst the spring wheat and the barley. The field workers devote their busy hours to carrying seed to the sower, turning dunghills in preparing manure for the potato and turnip crops, and continuing the barn work to supply litter for the stock yet confined in the steading, and to prepare the seed corn for the fields. The hedger now resumes his work of water tabling and scouring ditches, cutting down and breasting old hedges, and taking care to release the sprouting buds of the young quicks from the face of the hedge bank which he planted at the commencement and during fresh weather in winter. The steward is now on the alert, sees to the promotion of every operation, and entrusts the sowing of the crops to none but himself, except a tried hand, such as the skilful hedger.

The farmer himself now feels that he must be 'up and doing'; his mind becomes stored with plans for future execution. The business of the fields now requiring constant attendance, his mind as well as body becomes fatigued and, on taking the fireside after the labours of the day are over, seeks for rest and relaxation rather than mental toil.

Cows Calving, and Calves

The first great event in spring, on a farm of mixed husbandry, is the calving of the cows; from eight to ten weeks at this season is a period of great anxiety for the state of the cows; and, indeed, till her calving is safely over, the life of the most valuable cow is in jeopardy.

Signs that a cow is in calf

Cows may be ascertained to be in calf between the fifth and sixth months of their gestation. The calf may be felt by thrusting the points of the fingers against the flank of the cow, when a hard lump will bound against the abdomen, and the feeling will be communicated to the fingers. Or when a pailful or two of cold water is drank by the cow, the calf kicks, when a convulsive sort of motion may be observed in the flank, by looking at it from behind, and, if the open hand is then laid upon the space between the flank and udder, this motion may be most distinctly felt.

The exact time of a cow's calving should be known to the cattleman as well as by the farmer himself, for the time when she was served by the bull should have been marked down. Although this last circumstance is not a certain proof that the cow is in calf, yet if she passes the period when she should take the bull again without showing symptoms of season, it may be safely inferred that she became in calf at the last serving, from which date should then be calculated the period of gestation, or of reckoning, as it is called. A cow generally goes nine months, or 273 days, with calf, although the calving is not certain to a day.

About a fortnight before the time of reckoning, symptoms of calving indicate themselves in the cow. The loose skinny space between the vagina and udder becomes florid; the vagina becomes loose and flabby; the lower part of the abdomen rather contracts; the udder becomes larger, harder, more florid, hotter to the feel, and more tender-looking; the milk veins along the lower part of the abdomen become larger, and the coupling on each side of the rump bones looser; and when the couplings feel as if a separation had taken place of the parts there, the cow should be watched day and night, for at any hour afterwards the pains of calving may come upon her. From this period, the animal becomes easily excited and, on that account, should not be allowed to go out, or be disturbed in the byre. In some cases, these entire preparatory symptoms succeed each other rapidly, in others they follow slowly; and the latter is particularly the case with heifers with the first calf. These symptoms are called springing, and the heifers which exhibit them are springers.

A new-born shorthorn calf takes its first tentative steps

The process of calving

There are a few preparatory requisites that should be at hand when a cow is about to calve. Two or three rein ropes are useful, to fasten to the calf if necessary – a flat soft rope being the best form, but common rein ropes will answer. A mat or sheeting, to receive the calf upon in dropping from the cow, should she be inclined to stand on her feet when she calves. The cattleman should have a calf's crib, well littered. The shepherd should pare the nails of his hands close, in case he should have to introduce his arms into the cow to adjust parts; and he should supply himself with goose grease or hog's lard to smear his hands and arms.

All being thus prepared, and the byre door closed to keep all quiet, the cow should be attended to every moment. The symptoms of calving are thus exactly described by Skellett, as they occur in an easy and ordinary case. 'When the operation of calving actually begins,' he says, 'then signs of uneasiness and pain appear: a little elevation of the tail is the first mark; the animal shifts about from place to place, frequently getting up and lying down, not knowing what to do with herself. She continues some time in this state, till the natural throes or pains come on; and as these succeed each other in regular progress, the neck of the womb, or *os uteri*, gives way to the action of its bottom and of its other parts. By this action, the contents of the womb are pushed forward at every throe; the water bladder begins to show itself beyond the shape, and to extend itself till it becomes the size of a large bladder, containing several gallons; it then bursts, and its contents are discharged, consisting of the liquor amnios, in which, during gestation, the calf floats, and which now serves to lubricate the parts, and render the passage of the calf easier. After the discharge of the water, the body of the womb contracts rapidly upon the calf; in a few succeeding throes or pains, the head and feet of it, the presenting parts, are protruded externally beyond the shape. The body next descends; and in a few pains the delivery of the calf is complete.' The natural and easy calving now described is usually over in two hours, though sometimes it is protracted five or six, and even so long as 12 hours.

In regard to extracting twin calves, before rendering the cow any assistance, it is necessary to ascertain that the calves have made a proper presentation; that they are free of each other; that one member is not interlaced with, or presented at the same time with, the other. When they are quite separated, they each can be treated according to its own case.

On the extrusion of the calf, the first symptom it shows of life is a few gasps which set the lungs in play, and then it opens its eyes, and tries to shake its head, and sneer with its nose. The breathing is assisted if the viscid fluid is removed by the hand from the mouth and nostrils; and the thin membrane which envelops the body in the womb should now be removed, much torn as it has been in the process of parturition. The calf is then carried by two men, suspended by the legs, with the head held up between the forelegs, and the back downwards, to its comfortably littered crib, where we shall leave it for the present to attend still farther on its mother.

The afterbirth, or placenta, does not come away with the calf, a portion of it being suspended from the cow. It is got quit of by the cow by pressing, and, when the parturition has been natural and easy, it seldom remains with her longer than from one to seven hours. The usual custom is to throw the afterbirth upon the dunghill, or it is covered up with the litter; but it should not be allowed to lie so accessible to every dog and pig that may choose to dig it up.

After the byre has been cleansed of the impurities of calving, and a supply of fresh litter introduced, the cow, naturally feeling a strong thirst upon her from the exertion, should receive a warm drink. I don't know a better one than warm water, with a few handfuls of oatmeal stirred in it, and seasoned with a handful of salt, and this she will drink up greedily; but a pailful is enough at a time, and it may be renewed in a short time after, should she express a desire for it. This drink should be given her for two or three days after calving, in lieu of cold water, and mashes of boiled barley and gruel should be made for her, in lieu of cold turnips; but the oil cake should not be forgotten, as it acts at this critical period as an excellent emollient.

MILKING

It is desirable to milk the new-calved cow as soon as convenient for her, as, whether the labour has been difficult or easy, the withdrawal of milk affords relief. Milking is performed in two ways, stripping and nievling. Stripping consists of seizing the teat firmly near the root between the face of the thumb and the side of the forefinger, the length of the teat passing through the other fingers, and in milking, the hand passes down the entire length of the teat, causing the milk to flow out of its point in a forcible stream. The action is renewed by again quickly elevating the hand to the root of the teat. Both hands are employed at the operation, each having hold of a different teat, and moved alternately. The two nearest teats are first milked, and then the two farthest. Nievling, or handling, is done by grasping the teat round at its root with the forefinger like a hoop, assisted by the thumb, which lies horizontally over the forefinger, the rest of the teat being also seized by the other fingers. Milk is drawn by pressing upon the entire length of the teat in alternate jerks with the entire palm of the hand. Both hands being thus employed, are made to press alternately, but so quickly following each other, that the alternate streams of milk sounds to the ear like one forcibly continued stream. This continued stream is also produced by stripping. Stripping, then, is performed by pressing and passing certain fingers along the teat; nievling by the whole hand doubled, or fist, pressing the teat steadily at one place.

Milking should be done fast, to draw away the milk as quickly as possible, and it should be continued as long as there is a drop of milk to bring away. This is an issue which the dairymaid cannot too particularly attend to herself, or see it attended to by those who assist her. Old milk left in the receptacle of the teat soon changes into a curdy state, and the caseous matter not being at once removed by the next milking, is apt to irritate the lining membrane of the teat during the operation, especially when the teat is forcibly rubbed down between the finger and thumb in stripping. The consequence of this repeated irritation is the thickening of the lining membrane, which at length becomes so hardened as to close up the small orifice at the point of the teat. The hardened membrane may easily be felt from the outside of the teat, when the teat is said to be corded. After this the teat becomes deaf, and no more milk can afterwards be drawn from the quarter of the udder to which the corded teat is attached.

The milking pail is of various forms and of various materials. A pail of oak having thin staves bound together with bright iron hoops, with a handle formed by a stave projecting upwards, is convenient for milking in, and may be kept clean and sweet (see the illustration below). Of course, the pail cannot be milked full; but it should be as large as to contain all the milk that a single cow can give at a milking; because it is undesirable to rise from a cow before the milking is finished, or to exchange one dish for another while the milking is unfinished.

There is one side of a cow which is usually called the milking side – that is, the cow's left side – because, somehow, custom has established the practice of milking her from that side. Whichever side is selected, that should always be used, as cows are very sensitive to changes.

Cows, independently of their power to retain their milk in the udder, afford different degrees of pleasure in milking them, even in the quietest mood. Some yield their milk with a copious flow, with the gentlest handling that can be given; others require great exertion to draw the milk from them in streams no larger than threads. The udder of the former will be found to have a soft skin, and the teats short; that of the latter will have a thick skin, with long, tough teats. The one feels like velvet, the other no better than untanned leather.

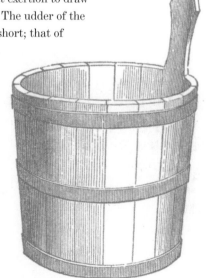

Milk pail

The milk that first comes from the cow after calving is of a thick consistency and yellow colour, and is called *biestings*. It has the same coagulable properties as the yolk and white of an egg beat up. After three or four days the biestings is followed by the milk. That which comes last, the *afterings* or *strappings*, as it is commonly called, is much the richer part of new milk, being not unlike cream. Being naturally thick, it is the more necessary to have it drawn clean away from the udder.

LOOKING AFTER THE CALF

Let us now attend to the young calf. The navel string should be examined that no blood be dropping from it, and that it is not in too raw a state. Inattention to this inspection may overlook the cause of the navel ill, the treatment of which is given below; and insignificant as this complaint is usually regarded, it carries off more calves than most breeders are aware of.

The first food which the calf receives consists of what its mother first yields after calving, namely, biestings. Being of the consistency of egg, it seems to be an appropriate food for the foetus just ushered into the world. On giving the calf its first feed by the hand it may be found to have gained its feet, or it may be content to lie still. In whichever position it is found, let it remain so, and let the dairymaid take a little biesting in a small dish – a handy form like a miniature milk pail, and of similar materials, will be found a convenient one – and let her put her left arm round the neck of the calf, and support its lower jaw with the palm of the hand, keeping its mouth a little elevated, and open, by introducing the thumb of the same hand into the side of its mouth. Then let her fill the hollow of her right hand with biestings, and pour it into the calf's mouth, introducing a finger or two into it for the calf to suck, when it will drink the liquid. Thus let her supply the calf, in handful after handful, as much as it is inclined to take. When it refuses to take more, the creature should be cleaned of the biesting that may have flown over. Sometimes, a calf being begun to be fed, when lying, attempts to get upon its feet, and, if able, let it do so, and rather assist it up than prevent it.

ONE MONTH OLD For the first month the calf usually has as much sweet milk warm from the cow as it can drink. It will be able to drink about 3 quarts [3.5 litres] at each meal, and in three meals a day, in the morning, noon and evening, it will consume 8 quarts [9 litres]. After the first month it gets its quantity of milk at only two meals, morning and evening. It is supported three months in all on milk, during which time it should have as much sweet milk as it can drink. Such feeding will be considered expensive, and no doubt it is, but there is no other way of bringing up a good calf.

The jelly from linseed is easily made by boiling good linseed in water and, while it is in a hot state pouring it in a vessel to cool, when it becomes a firm jelly, a proportion of which is taken every meal, and bruised down in a tubful of warm milk, and distributed to the calves. They are very fond of it, and in the third month of the calf's age, when it can drink a large quantity of liquid at a time, and during a day, it is an excellent food for them.

TWO MONTHS OLD When the air becomes mild as the season advances, and when the older calves attain the age of two months, they should be taken out of the cribs and put into the court during the day; and after a few days' endurance to the air, should be sheltered under the shed in the court at night, instead of being again put into the cribs. Some sweet hay should be offered them every day; as also, a few sheep slices of Swedish turnips to munch at. Such a change of food may have some effect on the constitution of the calves, causing costiveness [constipation] in some and looseness in others; but no harm will arise from these symptoms, if remedial measures are employed in time. Large lumps of chalk to lick at will be found serviceable in looseness. The shed of the court should be fitted up with small racks and mangers to contain the hay and turnips, and chalk. Should a very wet, snowy, stormy or cold day appear after the calves have been put into the court, they should be brought back to their cribs till the storm passes away.

THREE MONTHS OLD At three or four months old, according to the supply of milk and the ready state of the grass to receive them, the calves should be weaned in the order of seniority, due regard being at the same time had to their constitutional strength. If a calf has been always strong and healthy, it may be the sooner weaned from milk if there is grass to support it but, should it have ailed, or be naturally puny, it should be remembered that good sweet milk is the best remedy that can be administered to promote condition or recruit debility, and should be given with an unsparing but judicious hand. Calves, on being weaned, should not be deprived of milk at once; it should be lessened in quantity daily, and given at longer intervals by degrees, so as they may not be sensible of their loss when it is entirely withheld.

Sowing Spring Wheat

Wheat cannot be sown in spring in every sort of weather, and upon every variety of soil. Unless soil possesses a certain degree of firmness, that is, contains some clay, it is not considered adapted for the growth of wheat, at least it is more profitable to sow barley upon it; and unless the weather is as dry as to allow strong soil to be ploughed in the proper season, it is also more profitable to defer the wheat, and sow barley in due season. When wheat is sown in spring, it is usually after turnips, whether these have been entirely stripped from the land, or partly consumed on the ground by sheep.

Pickling wheat

Seed wheat should be pickled, that is, subjected to a preparation in a certain kind of liquor, before it is sown, in order to ensure it against the attack of a certain disease in the ensuing summer, called smut, which renders the crop comparatively worthless. Wheat is pickled in this way. For some days, say two or three weeks, let a tub be placed to receive a quantity of chamber ley [urine], and whenever ammonia is felt to be disengaging itself freely from the ley, it is ready for use. It is better that the effluvium be so strong as to smart the eyes, and water added to dilute the liquor, than that the ley be used fresh. This tub should be removed to the straw barn, as also the wheat to be pickled, and part of the floor swept clean, to be ready for the reception of the wheat. Let two baskets be provided capable of holding easily about half a bushel of wheat each, having handles raised upright on their rims. Pour the wheat into the baskets (*b*) from the sacks (*a*), and dip each basketful of wheat into the tub of ley (*c*) as far down as completely to cover the wheat, the upright handles of the baskets preventing the hands of the operator being immersed in the ley. After remaining in the liquor for

Apparatus for pickling wheat

two or three seconds, lift the basket up (*d*) to drip the surplus ley again into the tub, and then place it upon two sticks over an empty tub (*e*), to drip still more till another basketful is ready to be dripped. Then empty the dripped basket of its wheat on the floor (*g*), and as every basketful is emptied, let a person spread by riddling through a barn-wheat riddle, a little slaked caustic lime upon the wheat. Thus basketful after basketful of the wheat is pickled till it is all emptied on the floor, when the pickled and limed heap is turned over and over again till the whole mass appears to be uniform.

SETTING DOWN SACKS IN THE FIELD

There is some art in setting down sacks of seed corn on the field. It should be ascertained how many ridges of the field to be sown are contained in an acre, so that the sacks may be set down at so many ridges as each sack contains seed to sow the ground, allowing the specified quantity of seed to the acre.

The carrier of the seed is a fieldworker, and the instant the first sack of seed is set down, she proceeds to untie and fold back its mouth, and fill the rusky [seed basket], with seed, and carries the first quantity to the sower, who should be ready-sheeted awaiting her arrival on the head ridge at the side of the field. Her endeavour should be to supply him with such quantities of seed as will bring him in a line with the sack when he wants more.

Tools for the Job

Sowing sheet and hand-sowing corn

Rusky The rusky, or seed basket, is usually made of twisted straw laid in rows above each other, and fastened together by means of withes of willow. It is provided with a couple of handles of the same material, sufficient to admit the points of the fingers, and also a rim around the bottom, upon which it stands. It should be filled each time with just the quantity of seed, and no more, which the sower requires at one time. The mouth of the sack should be rolled round upon itself, that the seed may be easily and quickly taken out, for there is usually no time to lose when seed is sowing. The carrier should be careful not to spill the seed upon the ground on taking it out of the sack, otherwise a thick tuft of corn will grow uselessly upon the spot.

Sowing sheet The sower is habited in a peculiar manner; he puts on a sowing sheet (shown left). The most convenient form of sowing sheet is that of a semi-spheroid, having an opening at one side of its mouth large enough to allow the head and right arm to pass through, and by which it is suspended over the left shoulder. On distending its mouth with both hands, and on receiving the seed into it, the superfluous portion of the sheet is wound over the left arm and gathered into the left hand, by which it is held tightly, while the load of corn is securely supported by that part of it which passes over the left shoulder across the back and under the right arm. The right arm, which throws the seed, finds easy access to the corn from the comparatively loose side of the mouth of the sheet, between the left hand and the body of the sower. A square sheet, knotted together in three of its corners, and put on in a similar manner, is sometimes used as a sowing sheet; but one formed and sewed of the proper shape, and kept for the purpose, is a much better article. Linen sheeting makes an excellent material for a sowing sheet and, when washed at the end of the sowing season, will last many years.

BASKET A basket of wickerwork is very commonly used in England for sowing seed (shown below). It is suspended by girthing, fastened to the two loops shown on the rim of the basket, and passed either over the left shoulder and under the right arm, or round the back of the neck; and the left hand holds it steady by the head of the wooden stave shown on the other side of the basket.

English sowing basket

Broadcast sowing wheat seed

In sowing with one hand, the sower walks on the third and fourth furrow slices from the open furrow, which he keeps on his right hand. Taking always as much seed as he can grasp in his right hand, he stretches his arm out and back, with the clenched fingers looking forward, while the left foot is making an advance of a moderate step. When the arm has attained this position, the seed is begun to be cast, which is done with a quick and forcible thrust of the hand forward. At the first instant of the forward motion the forefinger and thumb are a little relaxed, by which some of the seed drops upon the furrow brow and in the open furrow, and while relaxing the fingers gradually, the back of the hand is turning upwards till the arm is stretched forward, when the fingers are all thrown open, with the back of the hand uppermost. The forward motion of the hand is accompanied by a corresponding forward advance of the right foot, which is planted on the ground the moment the hand has cast forward the most of the seed.

It is obvious that in sowing with the hand, the corn is scattered promiscuously and, in whatever arrangement it may come up, depends on the form of the ground, whether it had been ploughed in common furrows or in ribs; for, in the latter case, the corn comes up in rows or drills, the corn having fallen into the hollows of the ribs when sown; and in the former, broadcast, that is, equally over the surface of the ground.

Harrowing

The harrows follow the sowers, each sower keeping two pairs of harrows (shown right) employed when the land receives a double tine, that is, backwards and forwards on the same ground, which a breaking in of the seed should always be. On inclined ground, for the sake of the horses, that end of the field should be first sown which gives the horses the advantage of breaking the seed down hill. If the sowing commences at the top of the declination, the harrows start at once for the breaking in down the hill; but if it commences at the foot of the inclination, the harrows will have to go to the upper side of the field and begin there. Two pairs of harrows work best together, their united breadth covering the entire ridge, and lapping over the crown where the soil is thickest.

After the appointed piece of ground, whether a whole field or part of one, has been sown and broken in, the land is cross-harrowed a double tine, but as, in this, the ground is not confined within the breadths of ridges, the harrows cover as much of the ground as they can, and get over it in less time than in breaking in; and, besides, the second harrowing being easier for the horses, they can walk faster.

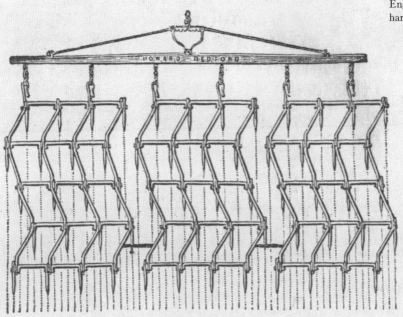

English iron harrows

SOWING OAT SEED

Beans and spring wheat are not sown upon every species of farm; the former being most profitable in deep, strong soils, and the latter is only to be commended after turnips, on land in good heart, situated in a favourable locality for climate, and the crop eaten off by sheep; but oats are sown on all sorts of farms, from the strongest clay to the lightest sand, and from the highest point to which arable culture has reached on moorland soil, to the bottom of the lowest valley on the richest deposits. The extensive breadth of its culture does not, however, imply that the oat is naturally suited to all soils and situations, for its fibrous and spreading root systems indicate a predilection for friable soils; but its general use as food among the agricultural population has caused its universal culture in Scotland, while its well-known ability to support the strength of horses, has induced its culture to be extended throughout the kingdom.

All the varieties of oats cultivated may be practically classed under three heads: the common, the improved and the Tartarian (see page 100). The common (or Potato) varieties include all those having a pyramidal spike, soft straw, long grains possessing a tendency to become awny [bearded], and which are late in reaching maturity. Among the named varieties are the following in common use: early and late Angus, Kildrummie, Blainslie, white Siberian, Cumberland, sandy and Dyock, which two last are recent varieties, and others. All common oats are sown on the inferior soils, and in the most elevated fields of farms, and the season for sowing them is the beginning of March.

The ploughed lea ground should be dry on the surface before it is sown, as otherwise it will not harrow kindly; but the proper dryness is to be distinguished from that arising from dry, hard frost. After the land is broken in with a double tine, it is harrowed across with a double tine, which cut across the furrow crests, and then along another double tine, and this quantity generally suffices. At the last harrowing the tines should be kept clean, and no stones should be allowed to be trailed along by the tines, to the manifest ribbing of the surface.

The land, after the oat seed is sown, is always water-furrowed in every open furrow and it should also be rolled according to circumstances; that is, the young braird [new shoots] on strong land being retarded in its growth, when the earth is encrusted by rain after rolling, it is safe to leave the rolling of such land until the end of spring, when the crop has made a little progress, and when the weather is usually dry. Light, friable land should be rolled immediately after the seed is sown and harrowed, if there is time to do it; but the rolling of one field should cause no delay to the sowing of others in dry weather. There will be plenty of time to roll the ground after the oat seed and other urgent operations at this season are finished.

Sowing Barley Seed

The walking on soft ground in sowing barley is attended with considerable fatigue, and as short steps are most suited for walking on soft ground, so small handfuls are best for grasping plump, slippery barley. Barley may be sown any time that is proper for spring wheat, and it may be sown as late as the end of May; but the earlier it can be sown, the better will the crop be in quality and uniformity, though the straw will be less. The average quantity of seed sown broadcast is three bushels to the imperial acre; when sown early less will suffice, and when late, more; because the later it is sown, there is less time for so quick a growing grain as barley to tiller and cover the ground.

There is no grain so easily affected by weather at seed time as barley: a dash of rain on strong land will cause the crop to be thin, many of the seeds not seemingly germinating at all, whilst others burst and cannot germinate; and in moist, warm weather the germination is certain and very rapid. Indeed, it has been observed, that unless barley germinate quickly the crop will always be thin.

The harrowing which barley land receives after being sown is less than oat land, a double tine being given in breaking in the seed, and a double tine across immediately after. Then the grass seeds are sown with the grass seed sowing machine, the land is harrowed a single tine with the light grass-seed harrows; water-furrowed (see page 133); and finished by immediate rolling. On strong soil, apt to encrust on the surface with drought, after rain the rolling precedes the sowing of the grass seeds, and the process is finished with the light grass-seed harrows; but on all kindly soils the other plan is best for keeping out drought, and giving a smoother surface for harvest work.

Effects of bad ploughing and sowing on seed

The subject of the condition in which seed is usually found deposited in the ground may be pursued a little farther, and the first conditions that strike one are bad ploughing and bad sowing.

In the diagram on page 128 (top), the furrow slices under a, b and c are all ill-ploughed, as you may easily perceive; and the soil being so arranged, of course the seed will be irregularly deposited, as seen at a, b and c in the second diagram down, where some are too deep, as under b, to germinate quickly, if at all, and others much nearer the surface, as at c. In these various positions in which the seed is placed, it is obvious that the plants springing from them will appear and grow irregularly (third diagram down), where the plants at b are much further advanced than those at a.

In regard to bad sowing, although the furrow slices are regular and well ploughed in (page 129, top) from c to d, yet the seed having been irregularly deposited on the surface, they have arranged themselves in one place too thick, as at $e\,e\,e$ (second diagram down), and too thin, as at $f\!f$ and the consequences are very visible in the position of the plants (page 128, bottom), where they are too thick at $g\,g\,g$, and too thin at $h\,h$.

Ill-ploughed irregular furrow slices

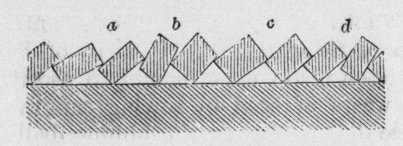

Irregular positions of seed on ill-ploughed furrows

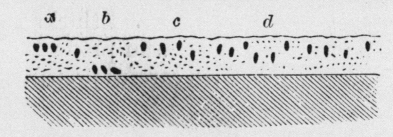

Irregular braird on ill-ploughed furrow

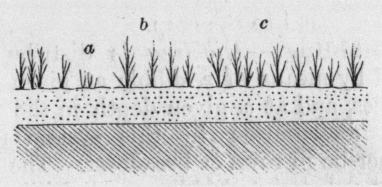

Irregular braird upon regular furrows

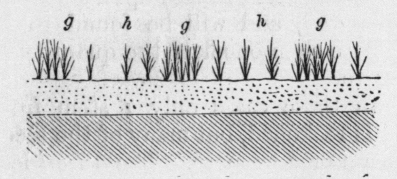

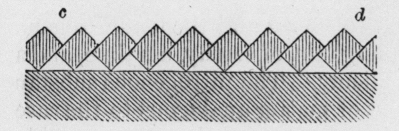

Well-ploughed, regular furrow slices

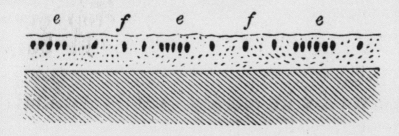

Positions of seeds on regular furrows

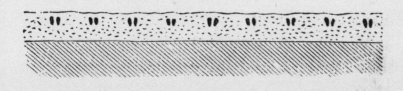

Regular depth of seed by drill-sowing

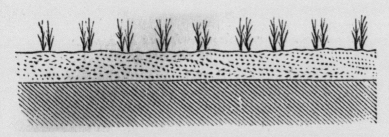

Regular braird from drill-sown seed

Sowing Grass Seed

Spring wheat following a green crop is always sown down with grass seeds, and the land is in a fit state to receive them when it is in the above condition of its harrowing.

Types of grass seed

The seeds usually sown amongst grain crops for the future grass crop consist usually of three kinds; annual red clover (*Trifolium pratense*), white or Dutch clover (*Trifolium repens*) and ryegrass (*Lolium perenne*). The proportions in which these seeds are sown depends on the rotation followed. If the grass is to continue in the ground only one year, a larger proportion of red clover is used than when it is to continue for two or more years. It is considered that 12lb [5.4kg] of clover seeds, and one bushel [3 UK gallons/36 litres] of ryegrass, is sufficient for an imperial acre. If the grass is to continue one year, the ryegrass should be the annual, and so called because it only affords a crop for one year, though by that time it has been two years in the ground – one with the crop in which it was sown, and one with the clover seeds – and though there is no botanical distinction between it and the true perennial ryegrass.

Mixing grass seed

The mode of mixing ryegrass and clover seed is this. The ryegrass seed is laid in a heap on the corn-barn floor, and the heap is made as flat on the top as to contain the clover seeds to be mixed with it. The red clover, being the larger-sized seed, is put on first, and spread over the top of the ryegrass; and then the white clover is poured over the red. The entire heap is then turned over by two barn shovels being made to meet

under the heap, and turned over behind the operator. This is done as
often as needed till the seeds, on being examined, appear well mixed.
Although the clover seed is so much heavier than the ryegrass, it does
not fall to the bottom of the heap, on account of its smallness, which
enables it to lie in the bosom of the ryegrass seed. The mixture is put
into sacks, and taken to, and set down upon the head ridge of the
ground sown with spring wheat.

Sowing grass seed

The sowing of grass seeds by hand is a simple process, though it
requires activity to do it well. Being sheeted as before and provided with
a carrier of the seed, the sower grasps the mixed seeds with two fingers
and the thumb, instead of the whole hand, and makes the cast and steps
exactly in the same manner as described in sowing corn. Clover and
ryegrass seeds being so very different in their form and weight, it is not
possible to cast them from the hand so that both shall alight on the
ground in the same manner. The sower, in fact, has little control over
the ryegrass seed, the least breath of wind making it go wherever it
may. His object simply is to cast the heavy clover seed equally over the
surface and, as it cannot easily be seen to alight on the ground, it is the
more necessary that he preserve the strictest regularity in his motions.

Grass-seed sowing machine

Such feats as just described cannot now be performed, because the
grass-seed sowing machine (illustrated overleaf) supersedes the
necessity of attempting it. It is a most perfect instrument for the
sowing of grass seeds, distributing the seeds with the utmost precision,
and to any amount, and scattering them so near the ground as to
render their sowing a matter independent of windy weather.

Its management is easy, when the ground is ploughed in individual
ridges; the horse which draws it walking in the open furrow, and the
machine reaching in length to the crown of the ridge on each side, sows
the width of a ridge at once, the length of the machine being made to

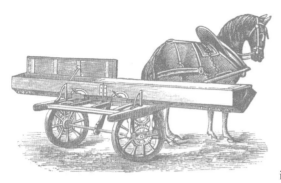

Broadcast sower

suit the breadth of the ridges adopted on the farm. The gearing is put out of action till the horse enters the open furrow from the head ridge, when it is put on, and it is again taken off when the machine reaches the opposite head ridge. The seed is supplied at one of the head ridges, and the head ridges are sown by themselves.

HARROWING

After the grass seeds are sown, the ground is harrowed to cover them in; and for this purpose lighter harrows are used than those for ordinary harrowing; and being light, are not infrequently provided with wings, to cover a whole ridge at a time, so that in following the sowing machine, the process connected with the grass seed sowing may be finished at once. There is some dexterity required in driving winged grass-seed harrows. It is not convenient to move them from one ridge to another immediately adjoining, as a part of the implement will have to turn almost upon a pivot; in doing which, unless conducted with great care injury is apt to be done to it. And, besides, it is a particularly awkward movement to hup [march] the horses with these harrows. The plan to avoid the inconvenience alluded to, is to hie the horses at the end of all the landings, and leave an intermediate unharrowed ridge at every turning, which will be greatly facilitated if the ploughman lifts up the near wing from the ground with a hooked stick when the turning is to commence, and lets it drop down again when it is finished.

ROLLING

The land may be rolled or not, according to circumstances, before the grass seeds are sown. If it is dry, even strong land would be the better, at this season, to be rolled, to reduce the clods before they become very hard, and to form a kindlier bed for the small seeds. Should it, however, be in a waxy state, between the wet and the dry, the rolling had better

Cast iron land roller

be deferred till afterwards. When it is in a proper state for rolling at the time of sowing the grass seeds, it should be rolled before the sowing, and, of course, before the harrowing of the grass seeds; because, were the land left with the smooth-rolled surface, and rain come after, and this succeeded by drought, which is not an infrequent state of the weather at this season, the smooth ground would become so battered and hardened, as to curb the wheat braird [shoots] considerably.

The roller is most conveniently formed of cast iron, and in two pieces, and mounted with shafts and framing, as shown above. The cast iron gives weight, which a roller should always have, and, being in two pieces, gives a facility to turn on little space. In driving it, the ploughman may sit on the front of the framing, if he wishes, and urge the horses with whip and reins. The framing sometimes supports a box, into which stones are placed to render the roller heavier, and this device may be necessary when reducing hard clods of clay in summer. Whether used for this purpose or not, the box is useful in carrying any stones that may be found on the land to either side of the field.

Water furrowing

The finishing process consists of water furrowing, that is, making a plough furrow in the open furrows, for the purpose of affording facilities to rainwater to flow off the surface of the land. It is executed with a common plough and one horse, or with a small double mouldboard plough and one horse; and in the execution, the plough obliterates the horse's footmarks. When the land is harrowed after the rolling, as in the case of heavy land, the water furrowing is done after the harrowing, and finishes the work of the field; but when rolling is the finishing operation, as in the case of light soils, the water furrowing is executed immediately after the harrowing and before the rolling. Water furrowing after rolling gives a very harsh-looking finish to a field.

Sowing Flax and Hemp

Flax

This plant was at one time pretty generally cultivated in this country, and the time is not far gone by when a given space of ground, of half an acre, was allowed for each hand to sow flax upon for his own linen, as a part of his wages. Finding this plant, as usually cultivated, in being permitted to ripen its seed, a severe one for the ground, and more probably because of foreign competition rendering its culture unprofitable, its cultivation here has almost fallen into desuetude. Of late the revival of its culture is spoken of.

Flax requires a deep, mellow soil, abounding in vegetable matter, removed both from strong clay and gravel; on the former of which the plant would be coarse, and on the latter, the crop scanty. Flax is cultivated after grass as well as after corn. In either case, the land should be early ploughed in autumn to afford time to receive the melioration of frost in winter. In spring the land cannot be worked too much by ploughing, harrowing and rolling, to render it as fine as garden mould. A clod-crushing roller has been seen to be the most ideal for such a purpose (shown right). The land should be finely harrowed before the seed is sown, which should be deposited at a very shallow depth in the soil, say ½ an inch [1.5cm]; and to improve the state of the soil still more, it should be rolled heavily and closely before being sown.

The seed may be sown any time from the middle of March to the first week of May. Being slippery, it is very difficult to sow by hand and, consequently, it is very apt to be laddered. The seed should be firmly seized by only two fingers and thumb, and sown in short, quick casts and, being dark-coloured, it may be observed to fall upon the rolled ground. The quantity of seed required for an imperial acre is from two to

two and a half bushels [16–20 UK gallons/72–91 litres] when a crop of fibre is desired; for a crop of linseed, less may be sown.

The flax crop cannot bear dung to be immediately applied to the land before it is sown, in case of its becoming too rank and coarse in the fibre; but a sprinkling of ten or 12 bushels [364–436 litres]of bone dust to the imperial acre in autumn or early spring, is said to be of service to the crop, in rendering its fibre finer upon land after a white crop.

Hemp

This plant is still less cultivated in this country than the flax. Hemp requires a deep, rich, mellow, alluvial, moist soil; and it should, moreover, be heavily manured on the stubble, with not less than 20 tons of dung per imperial acre. The working of the land is in every respect similar to that for flax.

Being a tender plant, very susceptible to frost, it should not be sown before the end of April. The quantity of seed sown is the same as for flax, namely, from two to two and a half bushels per imperial acre.

Crosskill's clod-crusher

Sowing Beans

The next field labour performed in spring, with the view of reaping a crop in the ensuing harvest, is the preparation of the land for the sowing of beans; not that the culture of this crop is general, for beans are not and cannot profitably be cultivated on every species of soil, they requiring the heavier and deeper class of soils, usually termed clays and clay loams.

The particular culture practised for raising beans is not dependent on the nature of the soil, but is meant to suit the nature of that plant's growth, and the state of the soil in reference to cleanliness. From the structure of the plant, which bears fruit pods on its stem near the ground as well as at the top, it is obvious that it should have both light and air; and its leaves being situated near the top, and its stem comparatively left bare of them, plenty of room near the ground is afforded to wild plants to grow in company with it. The plant possessing these properties, it is obvious that unless air is admitted to it, and opportunity afforded for removing weeds from it, it will not grow with that luxuriance which its nature would lead us to expect, if placed in favourable circumstances.

Now, there is only one way by which both these objects can be secured to the plant, which is, to place it in rows or drills. The air will then reach both sides of every row; and if the rows are placed so far asunder as to allow weeds to be removed as they grow up, the two objects of constitutional vigour of plant and cleanliness of soil will be attained; and, accordingly, beans are now usually sown in drills.

Beans within the Rotation System

The place which beans occupy in the rotation or course of crops, is not arbitrary; and no crop can be so treated that is cultivated in accordance with a predetermined system. They are considered a preparatory crop – though all crops that bear seed to perfection must be exhausters of the soil – because they allow the soil beside them to be worked during a considerable portion of the season; and, in practice, it is found they form an excellent preparation of the soil for wheat.

Harrowing

Suppose, then, you find the land in spring cloven down with gore [triangular] furrows, the first operation is to harrow down the furrow slices across the ridges, in doing which, the land being strong, and lying in a rough state, the harrows will take a firm hold of it, and tear it to

Newberry's one-rowed dibbling machine

pieces in a contrary direction from the one in which it had been cut by the plough in winter; and the immediate effect will be the filling up of the open furrow, and also of the gore furrows. The land, in fact, will be brought down nearly to a flat state. If the land, however, has become very much consolidated in winter, a cross harrowing will have little effect. It should, in that case, rather be harrowed along the furrows, and even that may prove of little service; seeing which, harrowing may altogether be desisted from. If the land is pretty dry, early as the season yet is, being most likely still in February, it can be harrowed well; but should it not be so dry as to bear the horses without much sinking, it had better be let alone for a few days, or even a week or two. Dry land and dry weather are both requisite for good harrowing; and in its turn harrowing exposes the land to drought.

THE BEAN DRILL

The bean drill or bean barrow is one of the simplest in its construction of that class of machines which I shall now have occasion to notice, and which are employed for depositing the various kinds of seeds in the soil in the drill system (shown on page 137). The bean drill is made in a form resembling a wheelbarrow, and hence its name, with a seed chest joined to form the bed frame of the barrow. The principal wheel, which is of very light construction and 18 inches [46cm] diameter, carries also a chain wheel upon its axle of the same diameter, and a pitch chain is stretched over the two wheels, by which means the progressive motion of the machine on the wheel gives motion to the seed cylinder on the axle. A spout, formed of sheet iron, is attached below to the bedframe, for the purpose of directing the seed to the furrow in which the machine is moving, and the legs are attached to the handles to prevent the latter from falling to the ground when the barrow is stopped.

Switching, Pruning and Water-tabling Thorn Hedges

Should a new line of thorn hedge have been required on your farm, and been proceeded with during winter, it will probably have engaged the hedger and his assistants all the time they could find available in the intervals between frosty, snowy and rainy weather. It being a matter of real importance to the farm to have every hedge planted on it in the best manner, the freshest state of the weather should be chosen to execute the work; and should the hedger have been prevented proceeding with the planting by an obstructive degree of frost, snow and rain, these same obstructions will have prevented him proceeding with his spring work among hedges, namely, the pruning of the hedges themselves, and the scouring of the ditches in connection with them. Before the season advances too far, these latter operations should not be neglected; and rather than not have time to do them in their proper season, the planting of the new hedge should be discontinued for the season.

Pruning

The first consideration which a hedge requires in spring is its pruning. The sap should either be entirely quiescent, or in a sluggish state, when pruning to any degree of severity is exercised on a hedge; and to secure this state of the plant the operation should be begun early in spring, but not till after the period of probably severe frost has passed away.

The pruning of hedges consists of two operations, switching and cutting down. Switching consists of lopping off straggling branches that grow more prominently from a hedge than the rest; and in doing this the extreme points of the other branches are also cut off. This operation is performed with the switching bill (right) which has a curved blade 9 inches [23cm] long and a helve 2 feet 3 inches [69cm] in length; and its weight altogether about 2lb [910g]. It feels light in hand, and is used with an upward stroke inclining backwards, something resembling a combination of cuts three and six of the cavalry sword exercise.

Hedgers have a strong predilection to use the switching bill. They will, without compunction, switch a young hedge at the end of the first year of its existence. No hedge ought to be touched with a knife until it has attained at least two years; because the great object to be attained by a new hedge is the enlargement of its roots, that they may search about freely for its support, and the only way it has of acquiring large roots is through its branches and leaves, which are the chief means of supporting the healthy functions of plants, or of even preserving them in life. Even beyond the age mentioned above, the pruning knife should be very sparingly used, until the young hedge has acquired the height sufficient for a fence; and not freely then, but only to remove superfluities of growth, and preserve equality in the size of the plants.

Let the plant have peace to grow till it has acquired a considerable degree of natural strength and let it be afterwards judiciously pruned, and I will give you the assurance of experience, that you will possess an excellent fence and a beautiful hedge in a much shorter time than the usual practice of hedgers will warrant.

Switching bill

Triangular form of thorn hedge

PRUNING OLD HEDGES

With regard to pruning hedges of older date, there are usually two forms of hedges found on a farm. One is the pointed or hog-mane shape, shown left, the other is a more natural form assumed by the plants, on having leave to shoot up their tops, whilst the lateral branches are switched off.

Of the two forms referred to, namely, the hog-mane and natural methods,

either may be adopted according to circumstances. Along both sides of a turnpike road, the low hog-mane is most advisable, to allow free circulation of air to the road. A height of from 4 feet [1.2m] to 6 feet [1.8m] will suffice for the purpose. The natural method is admirably adapted to afford shelter, and should therefore be reared against the stormy quarter of the farm; and, as pruning is attended with trouble and expense where hedges thrive luxuriantly, they may be allowed to grow up where they cannot do harm, such as on the tops of heights and in hollows. After having attained its natural height – which, in a hedge, may be 10 feet [3m], the thorn plant acquires thickness of stem and, if let alone, will continue to increase for many years; but while the stem becomes thicker, the plant changes its character, gradually forsaking the form of the hedge plant, and assuming the more natural form of the tree – enlarging its head by the lateral expansion of the upper branches, and increasing its stem by a natural pruning of the lower ones – every year thus rendering itself more and more unfit for a fence. In observing this natural tendency in hedges, the hedger should learn that the thorn plant is not in its natural position when placed as a hedge in a line along the side of a field; and, consequently, if he desires to retain it still as a fence, he must curb its tendency to become a tree. He may even make it resume its youthful habits, by well-timed pruning, and by the remarkably accommodating nature of its character.

Cutting down a hedge

The only sort of pruning applicable to such a case is cutting down the hedge, and there are two modes of doing this, one by leaving the stems and branches at a certain height from the ground; the other by cutting off all the branches and the stems to within a few inches of the ground. The first mode of cutting a hedge is called breasting it over, the second cutting it down. The former is done when the stems can be cut over with a light hedge bill; the latter requires a heavy bill or the hatchet.

The instrument with which a hedge is breasted over is called a *breasting knife* (shown overleaf) and it is very like the switching bill but the blade is somewhat shorter and stronger, and the bill is, of course, altogether heavier. It is used with a back-handed upward stroke.

Breasted thorn hedge on bank and ditch

Breasting knife

BREASTING

On determining to breast over a hedge, its stems should not be stronger than what a hedger can cut through by one hand with two or three strokes of the breasting knife. The hedger, on commencing this operation – using the knife with his right hand, and covering his left with a stout leather glove – stands on the hedge bank, in a line with the hedge, with his face outwards, that is, with his right hand to the hedge to be cut down. After cutting a few thorns at the end of the hedge, to make room for himself to stand upon, he commences cutting the principal stem or stems from one root, at about the height of his knee above the ground out of which they are growing.

In cutting, he uses the knife in this way. First making a firm cut upwards upon the stem, the knife may penetrate into its heart, or, if it is not much exceeding 1 inch [2.5cm] in diameter, may cut it through at one stroke; but the generality of the stems will require more than one stroke, though I have seen a hedger, of by no means great personal strength, cut through far thicker stems at a stroke than his appearance would indicate. But suppose the first stroke penetrates about an inch [2.5cm], the next one is given from above to meet the inner end of the first one, so that a wedge of the stem may be cut out; this flying off, the next stroke is given in the exact line of the first, and it will most probably sever the stem; but if not, a stroke at the furthest corner of the cut, and another at the nearest, will send the knife through. The cut stem will either drop down on end upon the ground behind the line of the hedge, or it will be kept suspended, by the interlacement of its branches amongst those of the plant beyond it. All the cuts made with a view to remove the wedge-shaped pieces are comparatively light; but all the upward cuts intended to sever the stem are given with force, and both kinds of strokes follow one another as fast as possible, until the stem is cut through. In renewing the strokes, the left hand is ready, the moment the knife is brought back, to receive its back between the fingers and thumb, as a rest. On the stem being severed, the hedger seizes its lower end with his gloved hand, and with the assistance of the knife in the other, pulls it asunder from the adjoining plant, and throws it endways either on the head ridge beyond the ditch beside him, or upon the head ridge of the field behind the hedge bank; whichever may be the place selected for a future dead hedge.

All the lateral branches growing from the stem are cut off in the same manner as far back as the top of the hedge bank, with an inclination corresponding to the slope of its face, so that the backmost branch preserves about the same height above the top of the hedge bank as the stem in front is above the hedge line. In the finished breasting sloping cuts are shown from *c* upwards (illustration left). Many of the branches will have been cut through with one stroke of the bill. The hedger proceeds in this manner until the entire hedge is cut down. Breasting is best suited to a comparatively young hedge, every branch and stem of which will soon be covered over with young twigs, which will form a close structure of vigorous stems.

Plashing

Hitherto the pruning and cutting have proceeded, on the supposition that the hedge cut down would make a sufficient fence when it grew up again; but this will not be the case if many of the stems are so far asunder as to leave gaps between them, even after the young twigs shall have grown. In such a state the pruned hedge will never constitute an efficient fence without farther assistance; and the mode in which assistance is rendered is by what is termed plashing, that is, laying down a strong and healthy stem across an opening which would otherwise form an irreparable gap in the hedge. When, on cutting down, the hedger meets with a gap which he sees will never be filled by the ordinary growth of the hedge, he leaves a healthy supple plant standing either beside or near it; and all such stems he leaves on the same side of a gap, that when plashed, they may all lie in the same direction.

After the hedge is all cut down, the hedger plashes down the stems he left standing, and this he does in the following manner. Commencing at the end of the hedge where he began to cut, he first prunes off all the branches and makes an upward cut in the stem near the ground, on the

Plashing and laying an old hedge

side opposite to the direction towards which the stem is to be plashed, but no deeper than is necessary to bend it to the proper position, which should be as near to and parallel with the ground as may be; because there is no use of plashing, unless the stems are laid so close to the ground as to fill up the gap from the bottom. The plashed stem is partly kept down by a snag on the neighbouring stem, or by twisting it before and behind others, or by a hooked stick driven into the ground near its point, and partly by a wedge thrust into the cut of its stem, and clayed up from rain and air. The plashed stem is cut over, of the length required for the particular gap. The breasting knife is used in plashing. Plashing may be seen above, where $e\ d$ is the first plashed stem, cut nearly through at e, and laid along near the ground, across the gap which extends beyond the cut; and where $b\ a$ is another stem passing across the large gap from b to c; k is the wedge of wood thrust into the cut of the plashed stem $b\ a$ to keep it down. The stem $b\ a$ extends beyond the immediate gap from b to c, if there is no means of fastening it down at c, and its end is brought in front of the pruned stem a; but should there be a means of fastening, then the stem should be cut off at c.

Dead Hedging

After the hedge has been pruned, as you have seen, let us now proceed to construct a dead hedge. A hedger and an assistant are necessary to construct it, and it is done in this manner. The assistant cuts the severed stems of thorns into pieces of about 3 feet [91cm] in length with a cutting bill or axe, according to the strength of the stems and, when these are very thick, they had better not be employed for this purpose, and the branches which they afford only lopped off and used. He lays one piece above another, until a bundle is formed that he can easily lift from the ground, taking care to add small twigs to it to thicken its appearance, and to compress it with his foot, which should be shod with a hedger's clog, that the pieces of the bundle may adhere to each other. He thus makes one bundle after another. The hedger meanwhile takes his station on the line chosen for the dead hedge to occupy, which is either immediately behind the hedge bank, or 1 foot [30cm] from the edge of the ditch in front of the hedge, according to the side on which it is intended to fence the hedge just cut down. The dead hedge, if placed behind the hedge, should not be set upon the top of the hedge bank, as cattle and horses would reach over it, and crop the shoots of the new-pruned hedge as they get up; it should occupy the foot of the hedge bank. It is a matter of some importance to construct a dead hedge so as not to be affected by the prevailing winds of the locality, otherwise it may be much torn and even upset by a high wind. For this reason its head should slope in the direction the dreaded wind blows. The first thing done by the hedger towards the formation of the dead hedge is to lay a spadeful of earth against the fence from which the dead hedge is to run (see illustration below).

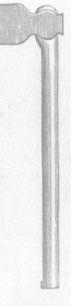

Hedger's axe

The trench made in the earth should be as large as easily to admit the lower end of the bundle of thorns. The first spadeful thus laid up forms a lean to the first bundle. When the hedger is ready with the trench, his assistant hands him a bundle with a long-shafted fork, which enables him to reach over the top of the breasted hedge. The hedger receives the bundle with his gloved hands, and places its butt end into the trench and, pushing it from him with his clogged foot, makes the head lie away from him. A tramp of earth is then raised with the spade, and placed against the butt end of the bundle. Thus bundle after bundle is set up firmly by the hedger, and after a few yards have been thus set up, he cuts in all straggling stems with the breasting knife, and chops the top and outside of the bundles into a neat form of dead hedge, having a perpendicular side and a flattish head.

Dead hedge of thorns

The Lambing of Ewes

The lambing season of Leicester and other heavy breeds of sheep, reared in the arable part of the country, commences about the middle of March, and continues for about the space of three weeks. There is no labour connected with the duties of the shepherd which puts his attention and skill to so severe a test as the lambing season; and a shepherd, whose unwearied attention and consummate skill become conspicuous, at that critical period of his flock's existence, is an invaluable servant to a stock farmer – his services, in fact, are worth far more than the amount of wages he receives; for such a man will save the value of his wages every year, in comparison with an unskilful shepherd, and especially in a precarious season, by so treating both the ewe and lamb, during the time, and for some time after the lambing season, that the lives of many are preserved that would otherwise have been lost.

Samuelson and Gardner's cylindrical turnip-cutter

Preparing for lambing

Before the season of lambing arrives, the shepherd should have a small field of 1 or 2 acres or a sheltered corner of a grass field of like size, conveniently situated as near the steading as possible, fenced round with nets or hurdles, and fitted up with sheds made of hurdles set up in the most sheltered part against a wall or hedge, and lined in the inside and comfortably roofed with straw. Such straw sheds form most comfortable places of refuge for ewes that may lamb in the night or that have lambed in the day, and require protection from frost, snow, rain or cold in the night, until the ewes are perfectly recovered from lambing, and the lambs sufficiently strong to bear the weather in the open field. The *turnip-cutter* (shown left) will be found on such occasions a very convenient instrument for cutting turnips in such turnip troughs (right) for the ewes in the paddock, or in small boxes for them in the shed. Common kale or curly greens are excellent food for ewes that have lambed, the nutritive matter of which is mucilaginous, wholly soluble in water, and beneficial in encouraging the necessary discharges of the ewe at the time of lambing.

Trough for turnip sheep-feeding

A large lantern which sheds plenty of light is an essential article of furniture at night to a shepherd. As foxes are apt to snatch away young lambs at night even close to the lambing houses, I have found an effectual preventive to their depredations in setting a *sheep net* in front of the lambing houses, leaving a sufficient space for a few ewes with their lambs taking up their lair within the net. When thus guarded, the foxes are afraid to enter the net, being apprehensive that it is set as a trap to ensnare them. Such an expedient is even more necessary in the corner of the field chosen for the lambing ground. A large lantern fixed on a stake within the lambing ground, and so placed as to throw light upon the whole ground, will be found a useful assistance to the shepherd in showing him the ewes that evince symptoms of lambing. A net and lantern are also good safeguards against foxes at night in the grass field, where the recovered ewes with their lambs should be gathered for the night.

The process of lambing

Being thus amply provided with the means of accommodation, the shepherd, whenever he observes the predisposing symptoms of lambing in as many ewes as he knows will lamb first – and these symptoms are, enlargement and reddening of the parts under the tail, and drooping of the flanks – he places them, of an afternoon, within the enclosed lambing ground in the paddock or field, as described above, and provides them with cut turnips.

The more immediate *symptoms of lambing* are when the ewe stretches herself frequently; separating herself from her companions; exhibiting restlessness by not remaining in one place for any length of time; lying down and rising up again, as if dissatisfied with the place; pawing the ground with a forefoot; bleating as if in quest of a lamb; and appearing fond of the lambs of other ewes. In a very few hours, or even shorter time, after the exhibition of these symptoms, the immediate

Border Leicester ewe and lamb

symptom of lambing is the expulsion of the bag of water from the vagina which, when observed, means the ewe should be narrowly watched, for the pains of labour may be expected to come on immediately. When these are felt by her, the ewe presses or forces with earnestness, changing one place or position for another, as if desirous of relief. Up to this time, not a hand should be put upon her, nor until the hoofs of the forefeet of the lamb, and its mouth lying upon them, are distinctly seen to present themselves in the passage.

When time has been given to observe that the ewe is not able to expel the lamb by her own exertions, it is the duty of the shepherd to render her assistance, before her strength fails by unavailing pressing. The exact moment for rendering assistance can only be known by experience; but it is necessary for a shepherd to know it, as there is no doubt that hasty parturition often induces inflammation, if not of the womb itself, at least of all the external parts.

ASSISTING IN LAMBING

When assistance should be rendered, the ewe is laid gently over upon the ground on her near or left side, and her head a little up the hill; and to prevent her being dragged on the ground when the lamb is being extracted, the shepherd places the heel of his left foot against the belly of the ewe, and kneels on his right knee on the ground across the body of the ewe, which lies between his heel and knee, with his knee pressing against her rump. Having both his hands free, and his face towards the tail of the ewe, he first proceeds to push out from him, with both hands, one leg of the lamb and then the other, as far as they will go; then seizing both legs firmly, above the fetlock joints, between the fingers of his left hand, he pushes them from him downwards from the ewe's back, with considerable force, whilst by pushing in the space between the tail of the ewe and the head of the lamb towards him, with the side of his right hand, he endeavours to slip the vulva of the ewe over the cantle of the lamb. These pushes are only given simultaneously with the pressing of the ewe, merely to assist her, and keep good what is obtained at each pressing, and not, as it were, to tear the lamb from her per force.

Whenever the head is cleared, the shepherd seizes the neck of the lamb behind the head with the right hand, and pulls out the body. The lamb is then placed at the ewe's head, for her to lick and recognize, which she will instantly do, if her labour has not been severe; but if it has, she will likely become sick, and be careless of the lamb as long as the sickness continues, which is evinced by quick breathing. If the pains have been very sharp, and this her first lamb, and she is not overcome by sickness, she may start to her feet, and run away from the lamb. The attempt should, of course, be prevented, and the tail of the lamb put into her mouth, to make her notice it.

While still lying on her side, her abdomen should be felt, to ascertain if there is another lamb to come; and if there is, the pains accompanying its passage may have been the cause of her carelessness for the first lamb; and if the second one is in a natural position, it will most probably by this time be showing itself in the passage, and if this be the case, the best plan is to take it away in the same manner as the first, and the ewe feeling the attempt, will at once assist on her part by pressing. The existence of a second lamb is worth attending to on another account, inasmuch as some ewes become so engrossed with the first lamb, that the pains attending the second are neglected by them, and they will indicate no signs of it for a time.

When a second is found in her, she must be watched, that whenever it comes into the passage, it may be taken away; but unless it actually makes its appearance there, it should not be attempted to be taken away. Should it not make its appearance in a reasonable time, it may be suspected that the lamb is either dead, or not in a natural position, and examination should be made by the fingers into the state of the case. A dead lamb is easily known by the feel, and should be extracted immediately, as it can afford no assistance of itself; but should the lamb be alive, it may be necessary to introduce the hand, to ascertain its position. Before the hand is introduced, it should be smeared over with goose grease. If the head is bent back, it must be placed straight, or if one leg or both be folded back, they must be brought forward, one by one, into the proper position.

After lambing

When lambing has taken place in the day, the ewe with her lambs are best at liberty within the enclosed area of the lambing ground, but in rain or snow, she should be taken under shelter to lamb, and kept there for some time until the weather prove better, or she be completely recovered from the effects of parturition. Should she lamb at night, it should be under cover, whatever may be the state of the weather. In the day, it matters not for lambs how cold the air is, provided it is dry.

The cleaning or placenta generally drops from the ewe in the course of a very short time, in many cases within a few minutes, after lambing. It should be carried away, and not allowed to lie upon the lambing ground. The lamb is fondly licked by the ewe at first, and during this process makes many fruitless attempts to gain its feet, but it is surprising how very soon after an easy birth it can stand; and the moment it does so, its first effort is to find out the teat, expressing its desire for it, by imitating the act of sucking with its lips and tongue, uttering a plaintive cry and wagging its long tail. It is considered a good sign of health when a lamb trembles after birth.

There are various obstacles to its finding the teat at first the long wool on the ewe's flank hides it, the wool on the udder interferes with it, and what is still more tantalizing to the anxious toper, the intense fondness of its own mother urges her to turn herself to fondle it with her mouth – uttering affectionate regards – but the motion has the effect of removing the teat, the very object of its solicitude. When at length a hold of what it wants is attained, it does not easily let it go until satisfied with a good drink. When a fond ewe has twin lambs, one can easily obtain the teat, while she is taken up in caressing the other. This is the usual conduct of strong lambs; and on once being filled with warm milk, progress rapidly to increasing strength, and are soon able to bear very rough weather. But lambs after a protracted labour, or the first lamb of young ewes, are so weakly at first as to be unable to reach the teat by their own strength, when they must be assisted, and the assistance is given in this way. Turning the ewe over upon her rump, the shepherd reclines her back against his left leg, which is bent, while he supports himself kneeling on the right one. Removing any wool from the udder by the hand, and which is all that is necessary, without clotting or doddering, as it is called in Scotland, he first presses the wax out of the teats, and then takes a lamb in each hand, by the neck, and opens the mouth with a finger, and applies the mouth to a teat, when the sucking proceeds with vigour.

The ewes are kept on the lambing ground till they have completely recovered from the effects of lambing and the lambs have become strong, and the ewes and lambs have become well acquainted with each other. The time required to accomplish this depends on the nature of the lambing, and the state of the weather. The ewes, with their lambs, are then put into a field of new grass, where the milk will flush upon the ewes, much to the advantage of the lambs. It is generally a troublesome matter to drive ewes with young lambs to any distance to a field, because of the ewes always turning round and bewildering the lambs. A dog more frequently irritates the ewes than assists the shepherd in this task.

I believe the best plan is to lead the flock instead of driving it, by carrying a single lamb, belonging to an old ewe, by the fore legs, which is the safest mode of carrying a lamb, and walking slowly with it before the ewe, while she will follow bleating close at the shepherd's heels, and the rest of the ewes will follow her of course. If the distance to the field is considerable, the decoy lamb should be set down to suck and rest. With plenty of food, and a safeguard of net and lantern at their lair at night, to keep off the foxes, the flock will not fail to thrive apace.

Shepherd's crook

SHEPHERD'S CROOK

This consists of a round rod of iron furnished at the point with a knob, that the animal may not be injured by a sharp point, and at the other end with a socket, which receives a long shaft of wood, 5 or 6 feet [1.5 or 1.8m] long, according to fancy (left). The hind leg is hooked in at a, from behind the sheep, and it fills up the narrower part beyond a, while passing along it until it reaches the loop, when the animal is caught by the hock and, when secured, its foot easily slips through the loop. Some caution is required in using the crook, for should the sheep give a sudden start forward to get away, the moment it feels the crook, the leg will be drawn forcibly through the narrow part, and strike the bone with such violence against the bend of the loop as to cause the animal considerable pain, and even occasion lameness for some days. On first embracing the leg, the crook should be drawn quickly towards you, so as to bring the bend of the loop against the leg as high up as the hock, before the sheep has time even to break off, and being secure, its struggles will cease the moment your hand seizes the leg.

Pasture for ewes and lambs

The state of the new grass fields occupied by ewes and lambs requires consideration. Ewes bite very close to the ground, and eat constantly as long as the lambs are with them; and as they are put on the new grass in the latter part of March, before vegetation is usually much advanced, they soon render the pasture bare when overstocked, and the weather is unfavourable to vegetation. In cold weather in spring, bitten grass soon becomes brown. Whenever the pasture is seen to fail, the ewes should be removed to another field, for if the plants are allowed to be bitten into the heart in the early part of the year, the greater portion of summer will elapse ere they will recover from the treatment. In steady growing weather there need be little apprehension of failure in the pasture. The sown pastures consisting chiefly of red clover and ryegrass, the clover is always acceptable to sheep; and in the early part of the season young shoots of rye grass are much relished by ewes. On removing the stock from the first to the second field, it is better to eat the first down as low as it safely can be for the plants, in order to *hain* it, that is, to leave it for at least a fortnight, to allow the young plants to spring again with vigour, and which they will do with a much closer bottom than if the field had been pastured for a longer time with fewer stock. Such a field eaten down to the end of May, or beginning of June, and allowed to spring afterwards in fine growing weather, will yield a much heavier crop of hay, than if it had not been depastured in spring at all. Although the whole breadth of young grass on a farm pastured lightly with ewes and lambs in the spring were to grow, as the season advances more rapidly than the ewes could keep it down, it will never produce the fine sweet fresh pasture which field after field will yield that has been eaten down in succession, and then entirely hained for a time.

TRAINING AND WORKING THE SHEPHERD'S DOG

The natural temper of the shepherd may be learned from the way in which he works his dog among sheep. When you observe an aged dog making a great noise, bustling about in an impatient manner, running fiercely at a sheep and turning him quickly, biting at his ears and legs, you may conclude, without hesitation, that the shepherd who owns him is a man of hasty temper. Most young dogs exhibit these characteristics naturally, and they generally overdo their work; and if you observe a shepherd allowing a young dog to take his own way, you may conclude that he also is a man who loses his temper with his flock. If you observe another shepherd allowing his dog, whether old or young, to take a range along the fences of a field, driving the sheep within his sight as if to gather them, you may be sure he is a lazy fellow, more ready to make his dog bring the sheep to him, than he to walk his rounds amongst them. Great harm may accrue to sheep by working dogs in these ways. Whenever sheep hear a dog bark that is accustomed to bound them every day, they will instantly start from their grazing, gather together and run to the farthest fence, and a good while will elapse ere they will settle again. And even when sheep are gathered, a dog of high travel, and that is allowed to run out, will drive them hither and thither, without an apparent object. This is a trick practised by lazy herds every morning when they first see their flock, and every evening before these take up their lair for the night, in order to count them more easily. When a dog is allowed to run far out, he gets beyond the control of the shepherd; and such a style of working among wether sheep, puts them past their feeding for a time; with ewes it is very apt to cause abortion;

Shepherd's dog or collie

and with lambs after they are weaned, it is apt to overheat them, and a considerable time will elapse before they recover their breath. Whenever a sorting takes place among the sheep, with such a dog they will be moved about far more than is necessary; and intimidated sheep, when worn into a corner, are far more liable to break off than those treated in a gentle manner.

A temperate herd works his dog in quite a different manner. He never disturbs his sheep when he takes his rounds amongst them at morning, noon and night, his dog following at his feet as if he had nothing to do, but ready to fulfil his duty, should any untoward circumstance require his services, such as breaking out of one field into another. When he gathers sheep for any purpose of sorting, or of catching particular ones, the gathering is made at a corner and to gain which he will give the sheep the least trouble, making the dog run out to the right and left, to cause the sheep to march quietly towards the spot; and after they are gathered, he makes the dog to understand that it is his chief duty to be on the alert, and with an occasional bark, prevent any of the sheep breaking away. When a sheep does break away and must be turned, he does not allow the dog to bite it, but only to bark and give a bound at its head, and thus turn it. In attempting to turn a black-faced wether [castrated ram] in this way, the dog runs a risk of receiving injury from its horns, and to avoid this, I have seen him seize the coarse wool of the buttock, and hang by it like a drag, until the sheep was turned round in the opposite direction, when he lets it go. In short, a temperate herd only lets his dog work when his services are actually required, he fulfilling his own duties faithfully, and only receiving assistance from his dog when the matter cannot be so well done by himself, and at no time will he allow his dog to go beyond the reach of his immediate control.

Dogs, when thus gently and cautiously trained, become very sagacious, and will visit every part of a field where sheep are most apt to stray, and where danger is most to be apprehended to befall them, such as a weak part of a fence, deep ditches or deep furrows into which sheep may possibly fall and lie awkward, that is, lie on the broad of their back and unable to get up, and they will assist to raise them up by seizing the wool at one side and pulling the sheep over upon its feet. Experienced dogs will not meddle with ewes having lambs at foot, nor with tups, being quite aware of their disposition to offer resistance. They also know full well when foxes are on the move, and give evident symptoms of uneasiness on their approach to the lambing ground. They also hear footsteps of strange persons and animals at a considerable distance at night, and announce their approach by unequivocal signs of displeasure, short of grumbling and barking, as if aware that those noisy signs would betray their own presence. A shepherd's dog is so incorruptible that he cannot be bribed, and will not permit even a known friend to touch him when entrusted with any piece of duty.

So far as my observation extends, I think there are two varieties of the shepherd's dog, one smooth, short-haired, generally black-coloured on the back, white on the belly, breast, feet and tip of the tail, with tan-coloured spots on the face and legs; the other is a larger and longer-bodied animal, having long hair of different colours, and long flowing tail. In their respective characters I conceive them to be very like the pointer and the setter. The small smooth kind, like the pointer, is very sagacious, slow, easily broken and trained and admirably suited to work in an enclosed and low country; the other, like the setter, is more swift, bold, ill to break, and requiring coercion, and therefore fitter for work on the hills.

Every shepherd's pup has a natural instinct for working among sheep, nevertheless they should always be trained with an old dog. Their ardent temperament requires subduing, and there is no more effectual means of doing so than keeping it in company with, and making it imitate the actions of, an experienced sober dog. A long string attached to the pup's neck, in the hands of the shepherd, will be found necessary to make it acquainted with the language employed to direct the various evolutions of the experienced dog while at work. With this contrivance it may be taught to '*hold away out by*', to '*come in*', to '*come in behind*', to '*lie down*', to '*be quiet*', to '*bark*', to '*get over the dyke or fence*', to '*wear*', that is, to intercept, to '*heel*', that is, to drive on, to '*kep*', that is, to prevent getting away; it will learn all these evolutions and many others in a short time, in imitation of its older companion and guide. It is supposed that the bitch is more acute than the dog, though the dog will bear the greater fatigue.

Turning Dunghills and Composts

Steel graip or dung fork

The ordinary mode of treating dunghills of farmyard manure is very simple. It is to spread every kind of straw used in litter, and every kind of dung derived from the various sorts of animals domiciled in the steading uniformly in layers, as it be supplied, over the area of the respective courts; to take this compound of straw and dung out of the courts at a proper period, and form it into a heap in the field where the manure derived from it shall be needed, with as much care as to mix the different ingredients of the heap together as they were in the courts, and to prevent the fermentation of the whole until the manure is used; and to turn this heap over in such a way, and at such a time, as the manure it contains shall be ready to be applied to the soil when wanted. The principle of this treatment is the simple one of commixing the various ingredients of straw and dung so intimately together, first in the courts, then in the dungheap, when led out, and lastly in the same dungheap when turned over to be duly fermented, as that the fermented manure shall be as uniform a compound as the nature of the materials of which it is composed will admit. And the result is, when the manure so treated is applied to the soil, it is found to be the most valuable of any known manure for every purpose of the farm.

My present purpose is to inform you how these heaps should be turned in order to bring them into the degree of fermentation best suited for making them into good manure; and the mode of actually applying that manure will be shown to you when we come to consider the culture of the potato and turnip crops.

Method of turning a dunghill

Laying these down as the rules by which dunghills should be turned, the mechanical part of the operation is done in the following manner. The people required to do this work are a man and a few fieldworkers, according to the size of the dunghills; and of this latter class, women are by far the best hands at turning a dunghill, because, each taking a smaller quantity of dung at a time upon a smaller graip (shown left), the dung is much more intimately mixed together than when men are occupied at this work, for they take large graipfuls, and merely lift them from one side of the trench they are working in to the other, without shaking each graipful loose, or scattering it to pieces.

The man's duty is to cut the dungheap into divisions of 3 feet [91cm] wide across its breadth with the dung spade in the manner described on page 105. When the edge of the dung spade becomes dull, it is sharpened with a scythe stone. The drier portions of the dung are put into the interior of the dunghill, and, when different sorts of dung are met with, they are intermingled in small graipfuls as intimately as possible. Each division of the dungheap is cut down and turned over to the ground before another one is entered on, and, when the ground is reached, the scattered straws, and earth that has been damped by any exudation [oozing] from the dungheap, are shovelled up either with the broad-mouthed shovel or the frying-pan shovel (right), and thrown into the interior. When straw ropes are met with, they should be cut into small pieces and scattered amongst the dampest portions of the dung. Though the dungheap is cut into parallel trenches, the dung from the top of one trench is not thrown into the bottom of the former, but rather thrown upon the breast of the turned dung, so as the turned dung may slope away from the workpeople. The utility of this form is, that when the dung is carting away it rises freely with the graip. When a dungheap is thus turned over, and its form preserved as it should be, it constitutes a parallelogram, and is a well-finished piece of work.

Frying-pan or lime shovel

Compost

The subject of composts, when followed out in all its bearings, is an extensive one – for there is not a single article of refuse on a farm, but what may form an ingredient of a compost, and be converted into a manure fit for one or more of the cultivated crops. At the same time there is great labour attending the formation of composts of every kind, as the materials cannot be collected together without horse labour; and in summer, when those materials are most abundant, the labours of the field are most important, and to devote then the time required to collect them, would be to sacrifice part of the time that should be occupied in indispensable field labour. I believe the most economical mode of forming composts is to collect materials at times when leisure offers for the purpose; and when they have accumulated in sufficient quantity in the space allotted for their use, they can be put together by the fieldworkers when not necessarily engaged in the field.

I made a point every year of making up a large compost heap of the quicken [organic matter] gathered from the fallow land, as it was preparing for the turnips – of the potato haulms as they were harrowed together – of the dried leaves in autumn, which would otherwise have blown about the lawn and shrubberies – and of any other refuse that could be collected together on the farm. These, with the assistance of a little fresh horse dung, and such water as the liquid manure tank, which was situated in the compost court, afforded, a compost was farmed every year, which assisted in extending the boundaries of the turnip crop; and if that portion of the crop was not always heavy, the greater or less proportion of the turnips eaten off by sheep, enabled it to produce its share of the succeeding corn crops and grass while, at the same time, the soil was thickened by the mould [mulch or organic-rich soil] reduced from the compost.

Any of the animals that fall by disease, when their carcasses are subdivided, and mixed in a large quantity of earth powdered with a little quicklime, will make a compost far superior to any of the preceding vegetable materials, for raising turnips, especially swedes.

The produce of privies, and pigeon's dung, as well as the dung of fowls from the henhouses, form excellent ingredients for putting into the tank of liquid manure to melt, and afterwards to add a compost.

Of late years sawdust, which was long considered a useless article, and which may be obtained in quantity on some farms where sawmills are at work, is now rendered useful in compost by being mixed with farmyard dung, fermented to a considerable degree of heat, and then subdued with water. Spent tanner's bark, when laid down for a time on the road around the steading and trampled underfoot and bruised by cartwheels, and then formed into a compost with either dung or lime, and allowed to stand for a considerable time, might be rendered into good manure for turnips. In the vicinity of villages where fish are cured and smoked for market, refuse of fish heads and guts, or liver and oil refuse, make an excellent compost with earth.

You thus see how endless is the subject of composts for manure; and it is obvious, from what has been said, that the kind of compost which you may make depends entirely on the nature of the materials which can be supplied in your immediate neighbourhood.

Planting Potatoes

The potato crop is cultivated on what is called the fallow break or division of the farm, being considered in the light of a green or ameliorating crop for the soil. Following a crop of grain, whose stubble is bare in autumn, the soil is ploughed early in the season, that it may receive all the advantages which a winter's sky can confer it in rendering it tender; and as potatoes effect a dry and easy soil, the piece of land intended for them may be ploughed and even partially cleaned in spring. Time for cleaning is very limited in spring, and on this account the cleanest portion of the fallow break should be chosen for the potatoes to occupy. The stubble will either have been cast in autumn, or clove down without a gore furrow, according as the soil is strong or free; and having been particularly provided with gaw cuts, to keep the land as dry as possible all winter, for a crop which requires early culture in spring, as potatoes do, it is probable that the land thus appropriated will be able to be cross ploughed, soon after the spring wheat and beans are sown, if either is cultivated on the farm, and if not, the cross ploughing for potatoes constitute the earliest spring work after the lea.

Flauchter spade at work

Preparing the Land for Potatoes

After the cross furrow, the land is thoroughly harrowed a double tine along the line of the furrow, and then a double tine across it, and any weeds that may have been brought to the surface by the harrowing are gathered off, along with any isolated stones that

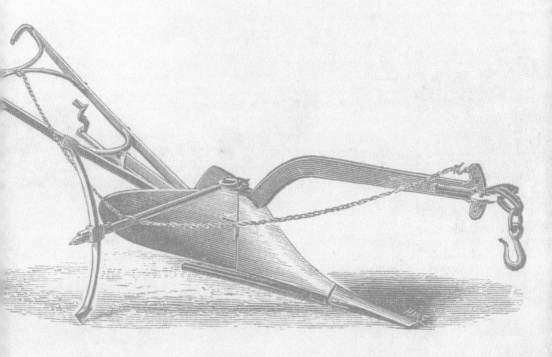

Scotch drill plough

would appear unseemly on the surface. If the land is clean, it will be ready for drilling, if not, it should receive another ploughing in the line of the ridges, that is, across the cross furrow, by being ridged up in casting, and then again harrowed a double tine along and across, and the weeds again gathered off. Should the surface be dry after the harrowing which succeeded the cross ploughing, and the weather appear not likely to continue dry, the *grubber* or *cultivator* will be a better implement to give a stirring to the soil than the plough, as it will still retain the dry surface uppermost, and it will also bring up to the surface any weeds that would entangle themselves about the implement.

The two-horse grubber is an excellent implement for stirring the soil when it has become somewhat solid by rain, or by lying untouched for a time. The time occupied in doing all this, as the weather will permit, may be about a month, that is, from the middle of March to the middle of April, when the potato crop should be actively preparing for planting. As the land cannot receive more ploughing in early spring than it should, to stir the land a little more, and make it still more tender, the drills first made for securing the manure of the potato crop should be set up in the double mode.

Preparing potato seed

While the land is preparing, and it will not be possible to prepare it with continuous labour, as the oat seed and the early part of the barley seed will have to be attended to, the seed of the potato crop must also be attended to, and this is more especially the duty of the fieldworkers. The potato pit is visited for the purpose (see page 288), the thatch and earth are removed and the potatoes are taken into a barn where they are prepared into sets to be planted.

The tubers are either planted whole, or cut into parts called sets. Whole potatoes create a great waste of seed, and when the sets are cut very small, the plants that arise from them are apt to be puny. The usual practice is to cut a middling-sized potato into two or three sets, according to the number of eyes it may contain, for unless there is at least one eye in every set, no plant will arise from it; and it is even precarious to have only one eye in a set, in case some accident should overtake it, or it may have lost its vitality. The sets should be cut with a sharp knife, be pretty large in size, and chiefly taken from the rose or crown end of the tuber. When fresh, the tubers cut crisp and exude a good deal of moisture, which, however, soon evaporates, and leaves the incised part of the set dry.

A very common practice is to heap the cut sets in a corner of the barn until they are to be planted; and perhaps, if they had been exposed to drought prior to this treatment, they might remain there uninjured, but if they are heaped up immediately after they have been cut and while quite moist, the probability is that those in the heart and near the bottom of the heap will ferment and evolve a considerable degree of heat, in which state it is doubtful that they will vegetate when planted, or at least that they will present so vigorous a plant as a non-heated set would originate. The safe plan, therefore, is to spread out the sets thinly on a floor, until they are fairly dried, when they may be put together somewhat thicker, but never on any account in heaps of 2 or 3 feet [60 or 91cm] in height.

Planting potatoes

Having drilled up as much land as will allow the planting to proceed without interruption, and having turned the dunghill in time to ferment the dung into a proper state for the crop, and having prepared the sets ready for planting, let us now proceed to the field, and see how operations should be conducted there, and in what manner they are best brought to a termination. The sets are shovelled either into sacks like corn, and placed in the field at convenient distances, or into the body of close carts, which are so placed on the head ridges so as to be accessible from all points.

TOOLS FOR THE JOB When the drills are short, the most convenient way to take the sets to the field is in a cart, as the distance to either head ridge is short; but when the drills are long, sacks are best suited for setting down here and there along the middle of the land. A small round willow basket, with a bow handle (illustrated below), should be provided for every person who is to plant the sets and, as a considerable number of hands are required for this operation, boys and girls may find employment at it, over and above the ordinary fieldworkers. A frying-pan shovel (see page 159) will be found a convenient instrument for taking the sets out of the cart into the baskets. Carts yoked to single horses take the dung from the dunghill to the drills. Graips are required to fill the carts with dung. Small graips are most convenient for spreading the dung in the drills; and a small common graip is used to divide the dung into each of the three drills as it falls into the middle drill from the cart. A *dung hawk* or *drag*, with two or three prongs, and about 5 feet [1.5m] long in the shaft, is used by the steward for pulling the dung out of the carts.

Potato hand basket

DUNGING Boys, girls or women, are required to lead the horse in each cart to and from the dunghill to the part of the field which is receiving the dung. The ploughmen, whose horses are employed in carting the dung, remain at the dunghill and, assisted by a woman or two, fill the carts with dung as they return empty. One man, the grieve or steward, hawks the dung out of the carts, and gives the land dung in such quantity as is determined on beforehand by the farmer. Three women spread the dung equally in the drills with the small graips, while a fourth goes before and divides it into each drill as it falls in heaps from the carts.

PLANTING Immediately that a part of three drills are dunged and the dung spread, the potato planters, after having filled their baskets or aprons with sets from the cart upon the head ridge, proceed to deposit the sets upon the dung along the drills, at about 8 or 9 inches [20 or 23cm] apart. Some women prefer to carry the sets in coarse aprons instead of baskets, because they are more convenient. As setting requires longer time than dung spreading, there should be two sets of planters to one set of spreaders, that is, six planters to four spreaders. One set of planters go in advance of the other till the latter comes up to the place where the former began, and then the second set goes in advance, and so one set after another goes in advance alternately, each set filling their baskets and aprons as they become empty, but all confining their labour to three drills at a time.

Whenever three drills are thus planted, the ploughman commences to split the first and cover in the dung and sets in the double way. The drills are split in the same way as they were set up; that is, in splitting the drills, turns over the first furrow upon the dung towards the planters and because of the first furrow being the largest, it should have complete freedom to cover the dung and sets.

Potatoes always receive a large dunging, they being in the first place a fallow crop, when the ground is entitled to be dunged and, in the next place, they are considered a scourging crop to the land, that is, taking much nourishment out of it, and returning little or nothing to it – yielding no straw but a few dry haulms, and the greatest proportion of the entire crop being sold and driven away from the farm. A large dunging to potatoes always seems great, for time is wanting to make the dung short, and, of course, to reduce its bulk in the dunghill. About 20 single horseloads, or 15 tons, to the imperial acre is as small a dunging as potatoes usually receive of farmyard dung.

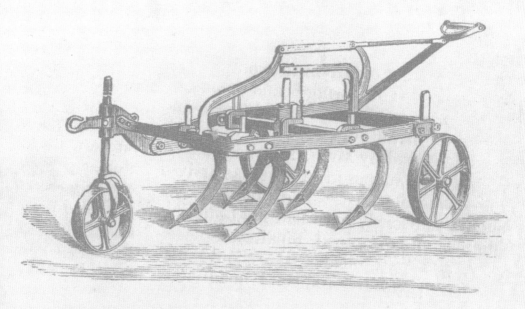

Broadshare cultivator or 'grubber'

THE GRUBBER

The grubber, or, as it is frequently styled, the cultivator (illustrated above), is a comparatively modern implement, and seems to have sprung from that more ancient prototype, the *brake harrow*, which it has now to a considerable extent superseded. It is in principle a member of the harrow tribe of implements; but while the harrow, or even the brake, acts superficially or to a very small depth in the soil, the grubber, from its greater weight, and from the form of its tines, penetrates the ground to any required depth. In respect to penetration, therefore, it approaches the plough, although its effects are simply to stir the soil, tearing up roots of plants at the same time, and it produces no effect towards turning it up like the plough.

It is, and will continue to be, an important implement, whether for the working of summer fallow, the preparing of rough or foul land for green fallow, or for the purifying of land that has been suffered to become so foul as to be incapable of producing a crop in the ordinary mode of cultivation. Soils in the last-mentioned condition, by passing through a course of grubbing that may be done in a few days or weeks at most, may be brought into a crop-bearing state. The grubber, like many other novelties, was, on its reintroduction in about 1811, held up as an implement capable of producing an entire revolution in the cultivation of the soil, and was spoken of as something above human invention, superseding even the plough; it has now settled down into its proper sphere of usefulness, ranking at least next to the plough.

BREAKING IN YOUNG DRAUGHT HORSES

Young draught horses are never broken in alone. They are most frequently yoked with an old steady horse at once into the harrows, accompanied with a few restrainers of reins and ropes, or an additional hand or two to assist the ploughman, to prevent any attempt at a run away; and, no doubt, when colts have been baltered [coupled with another horse] and led about from the time they were weaned by a steady, quiet-tempered man, they will soon submit to work, and become harmless in the course of a few short yokings.

Draught stallion

Breaking

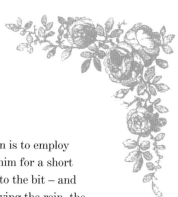

The easiest plan to make a draught colt work well soon is to employ a good horsebreaker to bridle, and handle, and lunge him for a short time – as long as is requisite to make his mouth yield to the bit – and then he will obey both voice and rein; and while employing the rein, the horsebreaker should be instructed to use the language that will be spoken to him while at work, the terms of which I have fully explained.

It is unnecessary to go through all the discipline of breaking in a draught colt, as is required in the case of a saddle horse, but a few preliminary steps are necessary, such as playing the bit in the mouth for two or three hours in the stable, twice or thrice a day, the colt standing in the reversed position in the stall, which has the double advantage of making the mouth yield to the bit, and of keeping up the horse's head.

After this discipline in the stable for two or three days, according as it is seen that the colt yields to the bit, with occasional walks out of the stable, he should be led out to walk two or three hours at a time by the nose rein, to learn to step out, and to acquire a good pace; and this is the most essential discipline for a draught horse. A short lunge or two backwards and forwards round a circle, on red land, will be useful, not to teach him to trot; but the trotting exercise will make him active, and sooner get the use of his legs in cases of difficulty. He should then be backed and, while guided by the reins, should be spoken to in the language he will be addressed in the yoke. After that he should be guided along a road with long double reins, while carrying the plough chains to accustom him to their noise and feel, and addressed in the appropriate language. Now all this discipline may be gone through in the course of a week, or eight or ten days, according to the disposition of the animal, the handling he may have received since he was a weaned foal, and the genius of the horsebreaker. The horsebreaker should groom the colt immediately after exercise, so the animal may become familiarized with the usages of the stable, and the degree of exercise given should be with a discrimination suited to the condition and physical strength of the animal.

Training with Harness

After the treatment and discipline received from the horsebreaker, the colt will be easily made to understand work. The sort of harness with which he is first invested is that of the plough, consisting of a bridle, collar and back band and chains. It is quite possible that the discipline received from the horsebreaker will make the colt suffer at once to be yoked with an old horse at the plough; but in case of accidents, and to err on the safe side, it is best to use precaution, even though it should be proved to have been unnecessary. The principal precaution is to attach the colt to a strong, steady horse, that will neither bite nor kick him, and be able to withstand the plunges the colt may choose to make. The attachment is made by a cart rope being first fastened round the girth of the old horse, and then passed round that of the colt, leaving as little space between their bodies as is required for ploughing; and to afford no liberty to advance or retire beyond a step or two before or behind the old horse. Beside the usual rein employed by the ploughman, the horsebreaker should have another in his hand from the colt's head.

Thus equipped in plough harness, the first yoking of the colt should be to an old cartwheel, placed on its dished face on ploughed land, furnished with a swing tree, which he should be made to draw, while the horse walks beside him; and in drawing this, the reins should be used, and the appropriate language spoken, that he may associate the changes of his motions with the accompanying sounds, and which are indicated by the reins while guiding him. I remark in passing, how curious it is for us to have adopted the Roman method of breaking young horses by the employment of the wheel, as set forth in the motto selected from Virgil. Should the colt offer to wheel round, the gentlest means should be used in putting him again in his proper position, as the start may have been made from fear, or from the tickling of a part of the harness. When a hind leg gets over a trace chain, the chain should be unhooked from the swing tree, and hooked on again after the colt has been put in his right position. Should he offer to rear or kick, from a disposition to break away, the old horse should be urged on to the walk, and be made to pull him along, while a smart tip of the whip will take the courage out of him. According as he evinces a disposition to go on quietly in the work, should the length of time be determined at which he should work at the wheel.

Once broken

When submissive, he should be yoked to the plough, for there is no species of work which calls forth the sympathy of horses to one another in so short a time as when working with this implement; and after a few landings, it will be seen that he will work with energy and good will, and then he should be kindly spoken to, encouraged, and even fondled. The probability is, that his desire for the draught may be evinced too keenly, but the pace of the old horse should be subdued, and the keenness mitigated by the rein and tug, which the short reins are called, that pass from the head of one horse to the collar of the other, and which, in this particular instance, is fastened to the rope round the girth of the old horse. It is interesting to the farmer to see his young horse put his shoulder to the first work he has ever tried with a spirit even beyond his strength; and while he continues at the work until his nostrils distend and flanks heave, his owner cannot help having a regard for him, heightened by a feeling of pity for the unconscious creature acquiring experience of work at which he is about to be doomed to toil for the remainder of his life.

The colt should be broken in to the cart as well as the plough. He is yoked into a single horse cart, but great care should be used on the first yoking, that he get no fright, by any strap rubbing against him, or the shafts falling upon him when raised up to allow of his being backed below them, for if frightened at the first yoking to a cart, a long time will elapse ere he will stand the yoking quietly. The horsebreaker should stand in the cart using double reins; and a rein should be held by a man walking first on each side of his head, and then at a little distance on the sides of the road. The chief danger is kicking, and thereby injuring the hocks against the front bar of the cart, to prevent which a rope should be placed across the top of the colt's rump, and fastened to the harness on the rump, and on each side to the shaft of the cart. There is little danger of his running away while all the reins are good.

Farm tip cart

Sows Farrowing or Littering

It should be so managed where there are more than one brood sow on a farm, to have one to bring forth pigs early in spring; but, at the same time, it should be borne in mind that young pigs are very susceptible of cold, and if exposed to it, though they may not actually die, their growth will be so stinted as to prevent them attaining to a large size, however fat they may be made. Even the most comfortable housing will not protect them from the influence of the external air, any more than certain constitutional temperaments can be rendered comfortable in any circumstance in spring, when under the influence of the east wind. From March to September may, perhaps, be considered as the period of the year when young pigs thrive best.

Whenever a brood sow shows symptoms of approaching parturition (labour); that is, when the vulva is observed to enlarge and become red, it is time to prepare the sty for her reception, for she will keep her reckoning not only to a day but to an hour.

The period of gestation of a sow is 112 days, or 16 weeks. The apartments meant to accommodate brood sows in the steading consist of an outer court 18 feet long by 8 feet broad [5.5 × 2.4m], enclosed by a door, and an inner apartment 8 feet by 6 [2.4 × 1.8m] roofed in. This is the usual form of a sty for sows, but others more convenient for overlooking the state of the sow and her pigs are when the outer court and inner apartment are placed under one roof, that is, in a roofed shed, or in a house which may be shut in by a door (see illustration on page 174). The litter allowed to a brood sow should be rather scanty and of short texture, such as chaff, short straw or dried leaves of trees, as young pigs are apt at first to be smothered or squeezed to death among long straw, when they get under it. When a sow has liberty before she is about to pig, she will carry straw in her mouth, and collect it in a heap in some retired corner of a shed, and bury herself amongst it, and the chance is, in such a case, that some of the pigs will be lain down upon unseen and smothered by the sow herself; when seen she will carefully push them aside with her snout before lying down.

Knowing the day of her reckoning, she should be attended to pretty frequently, not that she will probably require assistance in the act of parturition, like a cow or a ewe, but merely to see that all the pigs are safe, and to remove any one immediately that may be dead when pigged, or may have died in the pigging.

I do not know whether it is generally the case, but I have frequently observed that pigs leave the womb alternately in a reversed order; that is, they are projected by a head and breech presentation alternately, not uniformly so, but most frequently. There is no doubt, however, of the fact, that the first-born pigs are the strongest, and the last the smallest and weakest, in a large litter, such as upwards of 12, though the difference is less or scarcely observable in smaller litters of six or eight. The small, weak pigs are usually nicknamed *wrigs*, or *pock shakings*, and are scarcely worth bringing up; still, if there is a teat for them to lay hold of, they ought not to be destroyed. Sometimes there are more pigs littered than the sow has teats to give to each.

Extra pigs can, no doubt, be brought up by hand on cow's milk, but the last ones of a very large litter are usually so small and weak that they generally die off in the course of a day or two to the number of the teats. A young pig soon gets to its feet after birth, and as soon finds its way to the teat; but it can find no sustenance from it until the sow pleases, so that until the entire parturition is accomplished and the sow recovered from it, there is no chance of the pigs getting a suck. Many sows are very sick during parturition, and for some time after; so much so that the skin of their mouth becomes bleached and parched, and the breathing quick. To those unaccustomed to see a sow in that state, it would seem that she must die; but a little rest recovers her, and she betakes herself fondly to her young.

There is a peculiarity exhibited by young pigs, different from the young of other domesticated animals, in each

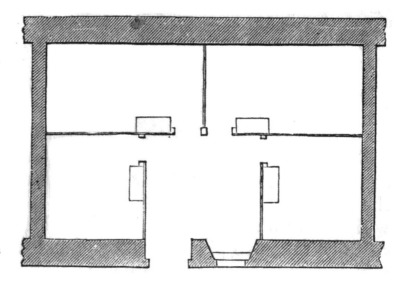

Sties for brood sows under one roof

choosing a teat for itself, and ever after keeping possession of it; and this compact, as it were, is faithfully maintained. Should there be one pig more than there are teats, it must take its chance of obtaining a teat when the rest are satisfied. It is generally observed that the pigs which are supported at the foremost teats become the strongest.

Pigs require to use coaxing before the sow will give them milk. They make loud entreaties, and rub the udder with their noses to induce her to lie down, which, when she does, every pig takes its own place right earnestly, and nuzzles away at the udder with the teat held in the mouth, whether situated in the upper or lower row. After a good while of this sort of preparation, the milk begins to flow on the sow emitting a fond-like grunting sound, during which the milk is drawn steadily and quietly till the pigs are all satisfied, and they not infrequently fall fast asleep with the teat in the mouth. Young pigs are lively, happy creatures, and fond of play as long as they are awake, but they are great sleepers. When a week old, their skins are clean, hair soft and silky, and with plump bodies and bright eyes, there are few more beautiful young animals to be seen about a farmyard. Those of a white colour look the most delicate and fine.

As to the food of the sow after she has recovered from parturition, which will be longer or shorter according to her constitutional temperament, she should get a warm drink, consisting of thinnish gruel of oatmeal and luke-warm water, and which serves the double purpose of meat and drink. If she is thirsty, which she is likely to be on recovery from sickness, the gruel may be again offered in a thinner state an hour or two afterwards. The ordinary food may consist of boiled potatoes, with a mixture of barleymeal, amongst water, administered at a stated hour at morning, noon and night, with such refuse as may occur from the farmhouse. This food will be found to support her well while nursing; and it should be borne in mind, that as long as she is nursing she should receive abundance of food if it is desired she should rear good pigs.

The swineherd looks on as his pigs enjoy the nut harvest of the beech wood

The Hatching of Fowls

Spring is the busy season of the feathered inhabitants of the farm. Thomson well describes the 'homely scene', which these happy creatures present while tending their young, and which might be seen at every farmstead in spring, were fowls cared for as they should be. Instead, however, of indulging in unavailing regrets, I shall endeavour, in as few words as the clear elucidation of the subject will admit of, to describe the mode of hatching and rearing every sort of fowl usually domesticated on a farm, and thereby show you that it is not so difficult nor so troublesome an affair as the practice which generally prevails would seem to indicate. Observation of the habits of domesticated birds, and punctual attention to their wants, are all that are required to produce and bring up plenty of excellent poultry on a farm.

Hens

As soon as the grass begins to grow in spring, so early will cared-for hens delight to wander into sheltered portions of pasture, in the sunshine, in front of a thorn hedge, and pick the tender blades, and devour the tiny worms, which the genial air may have warmed into life and activity. With such morsels of spring food, and in pleasant temperature, their combs will begin to redden, and their feathers assume a glossy hue; and even by February their chant may be heard, and which is the sure harbinger of the commencement of the laying season. By March a disposition to sit will be evinced by the early-laying hens; but every hen should not be allowed to sit; nor can any hen sit at her own discretion where the practice is, as should be, to gather the eggs every day as they are laid. It is in your own option, then, to select the hens you wish to sit to bring out chickens. Those selected should be of a quiet social disposition, not easily frightened, nor over careful of their brood, nor disposed to wander afar; and they should be large and full-feathered, to be able to cover their eggs well, and brood their young completely. Those which you may have observed to be good sitters and brooders should be chosen in preference to others; but it is proper to begin one young hen or so, every season, to sit for the first time.

The eggs intended to be set should be carefully selected. If those of a particular hen are desired to be hatched, they should, of course, be kept by themselves, and set after her laying time is finished. In selecting eggs, they should, in the first place, be quite fresh, that is, laid within a few days; they should be large, well-shaped, truly ovoidal; neither too thin nor too thick, but smooth in the shell; and their substance should almost entirely fill the shell, and should be perfectly uniform and translucent, when looked through at a candle, which is the best light for their examination.

A good cockerel is always on the lookout, alert to danger and protective of his hens

Either 11 or 13 eggs are placed under a hen; the former number, 11, is more likely to be successful in being entirely hatched than the latter, as few hens can cover in a sufficient manner so many as 13 large eggs.

As essential a matter as selecting the hens and eggs is the making of a proper nest for the sitting hen. This should consist of a circular hassock of soft straw ropes, or it may be a box or a basket. The object of this foundation is to raise the real nest sufficiently off the ground to keep it dry, and to give the nest such a hollow as that none of the eggs shall roll out by any mischance that may befall it. Either a box or a basket is a very convenient thing for the purpose, but in using either it will be requisite to stuff the corners, as well as the bottom, firmly with straw, that the eggs may not drop into the corners, or the young chicks, as they are hatched, fall into them. The nest itself should be made of soft short straw, sufficiently large to contain the hen, just sufficiently hollow to prevent the eggs rolling out, and sufficiently high above the floor to prevent any draught of air reaching the eggs.

Places should be chosen for placing the sitting hens in, for the henhouse, common to all the laying hens, will not answer, the perpetual commotion in it disturbing the sitting hens. A hatching house should contain one hen at a time, but as many may be accommodated in it as there are partitions to separate the one hen completely from the other, for hens are so jealous of each other, especially when sitting, that they will sometimes endeavour to take the nest and eggs from one another. Other places can be selected for sitting in, such as an outhouse, a loft, a spare room in the farmhouse or even the back kitchen. The hen selected for sitting having been accustomed to lay in the henhouse, or elsewhere, will feel annoyed at first on being transferred to her new quarters; she will have to be coaxed to it, and even after all may prove obstreperous, though exhibiting strong symptoms of clucking, in which case she must be dismissed and another chosen, rather than the risk be run of spoiling the whole hatching by her capricious conduct. A couple or so of old eggs should first be put into the nest, upon which she should be induced, by meat and water beside her, to sit for two or three days, to warm the nest thoroughly, before the eggs she is to hatch are placed under her. After she shows a disposition to sit, and the nest has become warm, the eggs are put into the nest, 11, as I said before, being quite enough, and the hen allowed to go upon them in her own way, and to manage the eggs as she chooses, which she will do with her bill and body, spreading herself out fully to cover all the eggs completely. The time chosen for setting the hen should be in the evening, when there is a natural desire for roosting and resting, and by next morning, by daybreak even, it will be found that she has taken to the nest contentedly.

Thus situated, the sitting hen should be looked at occasionally every day, and supplied with fresh food, corn and clean water. She will not consume much food during the time of incubation, which is three weeks. Every two or three days the dung, feathers, etc. about the nest or on the floor should be swept and carried away – the place should be kept clean and dry. In about three weeks a commotion among the eggs may be expected; and should the hen have proved a close sitter, and the weather mild, it is not unlikely that two or three chickens will be seen peeping out below her feathers before that period. The hen should not be disturbed during the time the chickens are leaving the eggs, or until they are all fairly out and dry. Any attempt to chip an egg infallibly kills the chick; and every attempt to remove pieces of a chipped egg causes the chick to bleed. A good plan is to give the chickens, when fairly out, a drink, by taking them one by one and dipping their bills in water. Meat is then set down to them on a flat plate, consisting of crumbled bread and oatmeal, and a flat dish of clean water. The hen's food consists of corn or boiled potatoes, and water. The chickens should be visited every three hours, and a variety of food presented, so as to induce them to eat it the more frequently and heartily, such as picks of hard porridge, crumbled boiled potatoes, rice, groats, pearl barley; taking care to have their food always fresh, and their water clean, however small the quantity they may consume. The hassock, or box, or basket, should now be removed, and the true nest set down on the floor, with a slope from it, that the chickens may have the means of walking up to the nest to be brooded at night. In the course of 24 hours after all the chickens are on foot, the hen will express a desire to go out, which she should be indulged in, if the weather is dry, and especially when the sun is out; but if it rain, she had better be kept within doors, unless there is a convenient shed near, in which she may remain with her brood for a short time. Visited every three hours during the day, and supplied with a change of food such as I have mentioned, and clean water, for about a fortnight, or until the feathers of the tails and the wings begin to sprout, chickens may then be considered out of danger. It is not expedient to set a number of hens at one time, but in succession every three weeks or a month; for a few chickens, ready in succession, are of greater value than a large number of the same age.

 Chickens go six weeks with their mother. A good hen that has brought out an early brood will become so fat while rearing them, that she will in time begin again to drop eggs, and of course again become a clucker, and may then be employed to bring out a late brood. Cock chicks, just out of the egg, are distinguished from hen chicks by their larger heads and stronger legs.

Turkeys

When a turkey hen is seen disposed to lay, a nest should be made for her in the hatching house. It may consist of the same materials as the hen's nest but, of course, of a larger size to suit the bird. A box or basket is an excellent thing, with every corner filled up. When once the turkey hen lays an egg, and a nest egg is placed in the nest, she will use it regularly every time she requires it, which will be once in about 30 hours. As the eggs are laid, they should be removed, and placed gently in a basket in the house, in a dry place, and turned with caution every day. When she has done laying, which may not be till she has laid 12 or 13 or even 15 eggs, she will be disposed to sit, when the eggs should be placed under her, to the number of 11 or 13, the former number being the most certain of succeeding, as a turkey cannot cover a greater number of her own eggs than a hen can of hers; and a brood of ten poults is an excellent hatching. Corn and water should be placed near the nest; but the turkey need not be confined within the apartment she occupies, though she is not disposed to wander, nor is she jealous of another one sitting in the same apartment with her. A turkey sits four weeks, and is proverbially a close sitter. During the incubation her food and water should be supplied to her fresh and clean daily, and all dung and dirt removed from her every two or three days.

When the poults are expected to make their appearance, the turkey should be frequently looked at, but not disturbed, until all the creatures are fairly hatched.

But before leaving the turkey for that night, the box or basket

A 'stag' or 'tom' turkey

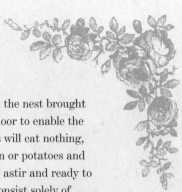

in which the nest is formed should be taken away, and the nest brought down upon the floor, with a sloping face towards the floor to enable the young poults to gain the nest. For 24 hours the poults will eat nothing, though the turkey herself should be provided with corn or potatoes and water. Next morning the young creatures will be quite astir and ready to eat food, which should now be given them. It should consist solely of hardboiled eggs, yolks and white shredded down very small, and put on a flat plate or small square board.

Let the poults be visited every three or four hours, supplied with hard-boiled egg and clean water. Let their food be removed after the poults are served, otherwise the turkey hen will devour it, for she is a keen feeder, and not so disinterested a bird as a hen. Let them remain two nights and a day in the house, and after that let them go into the open air and enjoy the sun and warmth, of which, it is hoped, there will be plenty in the month of May. In wet weather, however, they should either be confined to their house, or allowed to go into a shed. When the birds become strong and active, in the course of a few days let the turkey be placed in a coop on the green to curb her wandering propensity, until the poults can follow her, which they will be able to do after they have partaken of the hard-boiled egg for a fortnight.

Their meat can be put down on a plate on the green beyond the reach of the coop, and where the poults can help themselves, whilst the food of the turkey is placed within reach of the coop, consisting of corn, potatoes and water. After the feathers in the tails and wings of the poults are beginning to sprout, the egg may be gradually withdrawn, and hard-boiled picks of porridge, with a little sweet milk in the dish, to facilitate the swallowing of the porridge, should be given them at least four or five times a day at stated hours, and will make wholesome food, until the mother can provide insects and other natural food for them, which will form a desirable variety with the food given them, and then they will thrive apace, and grow amazingly fast as the weather becomes warm. Should the grass be damp, let the coop be placed on the gravel of the road or walk, as dampness is injurious to all young birds of the gallinaceous tribe. After the egg the poults are fond of a little shredded cress and mustard, and when at full liberty, they will pick the leaves of nettles with avidity. These predilections for ants, cress and nettles show that turkeys enjoy stimulating condiments with their food.

GEESE

The goose and gander cannot embrace but in water, and if the pond which they frequent be covered with ice, it should be broken to allow them to get to the water, and every egg requires a separate impregnation. An attentive observer will know when a goose is desirous of laying by her picking amongst straws and placing one on this side and one on that of her, as if making a nest. Whenever this is noticed, or an embrace on the water, a nest should be made for the goose to lay on in the hatching house, and to which she should have easy access, for she cannot jump up with the nimbleness of either a hen or a turkey, though her nest may be made in a box or basket, in the same manner as that of the hen, but of a size to suit the bird intended to occupy it. It is not proper to confine a goose a long time before laying her first egg; but when symptoms of being with egg are observed, she should be caught in the morning, when let out, and the lower portion of the soft part of the abdomen felt, where the egg may be easily ascertained to be in a state or not for being immediately laid, and if it feel hard, the goose should be shown her nest and confined to it until she lays in the course of the day, after which she is of course let out, the egg taken away and kept dry in a basket, and turned every day, until the whole are placed under her. Every other day after this the goose will visit her nest and lay an egg, and the number she may lay will seldom exceed 12, though 18 have been known to be laid; so, by the time she is done laying, it will be about the end of March.

After the goose has finished laying her eggs she will incline to sit, and then is the time for her to receive the eggs; and the best time for placing her upon them in the nest, as I have said before, is in the evening, that, by the arrival of the morning, the nest being warm and comfortable, the goose will have no inducement to leave it. The number of eggs given to be hatched should be 11, which is as many as a goose can conveniently and easily cover. The goose plucks the down off her body to furnish her nest with the means of increasing its heat; and one great use of the down is, that when she leaves her nest at any time it serves the double purpose of retaining the heat about the eggs and of preventing the external cold affecting them. A little clean water and a few oats are put beside a goose when she is sitting; but she will eat very little food all the time of her incubation. Let her go off whenever she pleases, and there is no fear but that she will return to her nest in time to retain the heat of the eggs; for she makes it a point to cover every egg with down before leaving the nest.

Let her go to the pond if she wills it, and wash herself in it and, depend upon it, she will not continue long there; she will be cooled and much refreshed by it. The feathers will not become wet; it is not their nature to become so; and after such a relaxation, which she much enjoys, she will sit the closer. Geese are liable to become costive [constipated] while sitting, and to counteract this tendency, they should be supplied now and then with a little boiled potato in a dry state, which they will relish much, and feel much the better for; and, indeed, every fowl, while sitting, should receive a little of this useful succedaneum [substitute] at that peculiar juncture. The gander usually takes up with one mate, but if there are only two geese, he will pay attention to both; and his regard for his mate is so strong, that he will remain at the door of the hatching house like a watchdog, guarding her from every danger, and ready to attack all and sundry who approach her sanctuary. At the end of a calendar month the eggs may be expected to be hatched; and during this process the goose should be left undisturbed, but not unobserved. After the goslings are all fairly out of the shell, and before they are even dry, they may be taken in a basket with straw to a sheltered dry spot in a grass field, the goose carried by the wings, and the gander will follow the goslings' soft whistle. Here they may remain for an hour or two, provided the sun shines, for in sunshine goslings will pick up more strength in one hour than all the brooding they can receive even from their mother for a day. The goose will rest on the grass, the goslings will endeavour to balance themselves on their feet to pluck it, and the gander will proudly protect the whole. Water should be placed beside them to drink. Should the sky overcast, and rain likely to fall, the goslings should be collected, and they and goose carried instantly to their nest; for if goslings get their backs wetted with rain or snow in the first two days of their existence, they will lose the use of their legs, never recover their strength, and inevitably die. Should the weather be wet, a sod of good grass should be cut and placed within their house, with a shallow plate of water. In setting down a common plate to goslings, it should be prevented from upsetting, as they are apt to put their foot upon its edge, and spill the water.

After two days' strength, and especially in sunny weather, the goslings may venture to the pond to swim; but the horse pond is a rather dangerous place for them as yet, so many creatures frequenting it. Some water, or a pond, in a grass field, would be a better place for them. For the first two or three days after goslings go about, they should be particularly observed, for should they in that time happen to fall upon their backs, or even into a hardened hoof print of a horse, they cannot recover their legs, will be deserted by their dam and the rest of the flock and will perish. After three or four days, however, in dry sunny weather, and on good grass, they will become strong, grow fast, and be past all danger. It is surprising how rapidly a young gosling grows in the first month of its life. After that time they begin to tire of grass, and go in search of other food; and this is the time to supply them daily with a few good oats, if you wish to have a flock of fine birds by Michaelmas; any other grain will answer the purpose, as rice and Indian corn, let it be but corn, though oats are their favourite food. Even ordinary light corn will be better for them than none; and if they get corn only till harvest, they will have passed their most growing period of their life, and will then be able to shift for themselves, first in the stack yard and afterwards in the stubbles. The sex of the gosling may be easily ascertained after the feathers begin to sprout, the ganders being white and strong in the leg, head and neck, while the geese are grey and gentler-looking. Goslings go with their parents for an indefinite length of time.

Ducks

It is customary to place duck eggs under hens, owing, I believe, to the great difficulty of making a duck take to a nest which she has not herself made. Hens make pretty good foster mothers to ducklings, though, in becoming so, the task is imposed upon them of a week's longer sitting than is natural to them. Still the entire production of ducklings on a farm should not be left to the chance of ducks setting themselves on eggs, for they are proverbially careless of where they deposit their eggs, and on that account hens must be employed to hatch at least a few broods.

As in the case of her own eggs, a hen can only cover 11 duck eggs with ease, and, of course, she requires the same treatment when sitting on them as she receives with her own eggs. A calendar month is required to bring out ducklings; and the hen should be left undisturbed until all the brood comes out. Ducklings should be kept from water for a couple of days, until their navel string is healed; and the food which they receive should be of a soft nature, quite the opposite of that given to turkey poults, such as bits of oatmeal porridge, boiled potatoes, bread steeped in water, barley-meal brose, and clean water to drink in a flat dish in which they cannot swim. On this treatment, three or four times at least every day, they will thrive apace, and become soon fledged in the body, when they are fit for use; but their quill feathers do not appear for some time after.

With regard to the means of supplying young fowls on a farm, two turkey hens and two geese will rear as many turkeys and geese as will be required by a large family; three or four broods of ducks will suffice; a brood or broods of chickens may be brought out from March to November; and as to pigeons, it is the farmer's own fault if he has not a supply of them from March to December. Those who do not care for the flesh of fowls may send their poultry to market; and those who breed for the market should provide a person well versed in rearing poultry to undertake the duty.

Hens are best used to hatch duck eggs

Summer

'The season now is all delight,
Sweet smile the passing hours,
And summer's pleasures, at their height,
Are sweet as are her flowers;
The purple morning waken'd soon,
The mid-day's gleaming din,
Grey evening with her silver moon,
Are sweet to mingle in.

How sweet the fanning breeze is felt
Breath'd through the dancing boughs;
How sweet the rural voices melt
From distant sheep and cows.
The lovely green of wood and hill,
The hummings in the air,
Serenely in my breast instil,
The rapture reigning there.'

CLARE

NTRODUCTION

I have represented Winter, in the agricultural sense, as the season of dormancy, in which everything remains in a state of quiescence. In the same sense I have said that Spring is the season of restoration to life, in which everything again stirs and becomes active. Summer, on the same principle, is the season of progress, in which nothing is begun or ended – none of the greater operations of the field are either commenced or terminated, but only advanced a step towards the consummation of all things in Autumn; and, therefore, the mere advancement of the greater operations involves no change of principle, whilst the smaller ones present so varied an aspect as to excite considerable interest.

The first operation which calls for the ploughman's attention in summer is the turnip land, which is now drilled up, dunged and sown. The culture of the turnip is a most important and stirring operation, affording much interesting work in singling [thinning] and hoeing the plants for the greater part of the season. In the height of summer, young stock luxuriate on the riches of the pasture field, while forage plants, consisting of vetches, rape, or broad clover, are allowed to grow until the general season of want, between the failure of pasture and the premature consumption of turnip. Before stock take possession of the pasture fields, the hedger makes it a point to put the fences in a complete state of repair, and to second his exertions, the carpenter and smith make the field gates secure for the season.

Fattened stock are seldom allowed to taste the pasture, they being disposed of off the turnips to the butcher or dealer. The fat cattle are almost always then sold. Young cattle and cows are sent to the grazing field of the farm, though turnip sheep are not infrequently retained on grass until the fleece is clipped from their backs, and after that they also are disposed of. The separation of ewe and lamb is now effected; and the respective marks of age, sex and ownership are put on each. Horses now live entirely another sort of life, being transferred from the confinement of the collar in the stable to the perfect liberty of the field, and heartily do they enjoy themselves there. The brood mare now brings forth her foal, and receives immunity from labour for a time.

Haymaking is represented by poets as a scene of unalloyed pleasure. No doubt lads and lasses are then as merry and chirping as grasshoppers but, nevertheless, in spite of buoyant spirits, haymaking, in sober truth, is a labour of much heat and great toil – the constant use of the hay rake and pitchfork, in hot weather, being no sinecure. Early as the season is, preparations are made in summer for the next year's crop. The bare fallow is worked and dunged, and it may be limed too, in readiness for the seed in autumn.

Summer is of all others the season in which the farmer most seriously makes his attacks on those spoilers of his clean fields, and contaminators of the samples of his grain – the weeds. Whether in stocked pastures, upon tilled ground, along drills of green crops, amongst growing corn or in hedges, young and old, weeds are daily exterminated, and the extermination is most effectually accomplished by the minute and painstaking exertions of female fieldworkers. For these purposes they are provided with appropriate cleaning instruments. This is the season, too, in which his stock and crops are sometimes seriously affected by the attacks of insects. Where building stones are plentiful, and the risk great from the overflowings of rivulets in winter, summer is also the season for the erection of stone dykes as fences between fields, and of embankments along the margins of rivers. The former afford a substantial fence at once, the latter form insuperable barriers against an element powerful alike whether exerted for or against man's operations.

Every operation requires constant attention in summer, for the season being active in its influences, farmers must then put forth their energies to meet its rapid effects, whether these tend to forward or retard his efforts. The long hours of a summer day, of which at least ten are spent in the fields – the ordinary high temperature of the air, which suffuses the body of the working man in constant perspiration – and the fatiguing nature of all field work in summer, bear hard as well on the mental as the physical energies of the labourer, and cause him to seek for rest at a comparatively early hour of the evening. None but those who have experienced the fatigue of working in the fields in hot weather, and for long hours, can truly appreciate the luxury of rest.

SOWING TURNIPS, MANGELWURZEL, RAPE, CARROTS AND PARSNIPS

The first great field operation in summer is the completion of the preparation of the soil for the sowing of the turnip crop. This crop commences the rotation of crops, is a substitute of bare fallowing, and is of the same nature, as regards the amelioration and working of the soil, as the potato crop, and therefore admits of the soil being manured; and, indeed, on account of all these properties, it is regarded and denominated a fallow crop. Being thus a renovator of the condition of the soil, the turnip crop necessarily succeeds the crop which terminates the rotation, and beyond which the exhaustion of the soil is not permitted; and, being a fallow crop, the preparation of the soil for it requires much labour, and should therefore be begun as early as the breaking up of the stubble in the beginning of winter.

Preparing the land

From the cross ploughing to the drilling of the land for the reception of the dung, the turnip culture is exactly the same as for potatoes, with perhaps the exception that, as there is more time for working and cleaning turnip land, it receives one or more ploughings or stirrings with the grubber than the potato land; and in this cleansing process the grubber will be found a most efficient implement, and will save a ploughing, while it keeps the upper soil uppermost and in a fine, loose state. When turnip land is manured with farmyard dung, the drilling is best and most expeditiously done in the single mode, but as the drills have to be kept in exact proportions, for the sake of the better operation of the sowing machine that is to follow, the best ploughman should be desired to make them.

So far the culture of potatoes and turnips correspond, but after this point a considerable difference ensues, which arises from the difference in the nature of the seed. After the soil on the top of the drills has become a little browned with the sun the turnip-sowing machine, which sows two rows at a time is then used, with one horse, for sowing the seed (see the illustration below).

Sowing

The quantity of seed sown need not exceed 3lb [1.4kg] to the English acre, nor should the quantity be much less, as thick sowing ensures a quick braird [sprouting] of the turnip plant, and the seed is not a costly article, being usually from 9d. to 1s. per lb. Fortunately, that the land may receive its due labour, the different kinds of turnips cultivated require to be sown at different times. Swedes, for instance, should be sown by the 15th of May at latest and, if the land is ready to receive the seed by the 10th, so much the better. Swedes will grow on any kind of sod, except perhaps what is in a state of pure peat; but they grow best in rich alluvial sandy loam – best, because largest, and in that state this particular turnip is firmest too. The yellow turnip follows the swedes, and then the white, which may be sown any time in June. In England, white turnips are sown as late as July, because, if sown earlier, they would come too soon to maturity.

Turnip drill sower

The established manures for raising turnips are farmyard dung, street manure in the neighbourhood of towns and bone dust. There are many other substances which have been recommended for the same purpose, such as guano, animalized carbon, etc.; but as they are only of comparatively recent introduction, and cannot be said to have yet established their characters, I shall decline entertaining their pretensions here.

SINGLING

The young turnip plants may be expected to make their appearance above ground in the course of eight or ten days at soonest, and later if the weather is unfavourable to vegetation. When the plants have attained about 3 inches [7.5cm] in length, it is time to prepare for their being singled, that is, thinned out singly to determinate distances. The first operation in preparation of the process of singling is passing the *horse hoe* between the intervals of the rows of the turnip plants. This implement, assuming different forms, is drawn by a single horse, and guided by a ploughman, and is figured and described below.

The object of using this implement at this time is partly the removal of any young weeds that may have shown themselves between the drills, but chiefly to pare away a little of the earth from each side of the drills in order to afford facility to the hand hoe in singling out the plants. The *improved hand hoe* (shown on page 194) used for singling turnips consists of an iron plate faced with steel, 7 inches [18cm] in length and 4 inches [10cm] in breadth, a swan neck which has an eye attached to its end to receive the shaft, usually made of fir, to make the

Turnip horse hoe

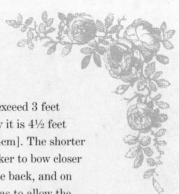

implement as light as possible. The shaft should not exceed 3 feet [91cm] in length, though in some parts of the country it is 4½ feet [1.4m], whilst in others it is as short as 33 inches [84cm]. The shorter it is the better for the work, as it enables the fieldworker to bow closer to the ground; but this position is very severe upon the back, and on this account the shaft is in some places made so long as to allow the worker to stand nearly upright, in which position, however, the eye and hand being removed a considerable distance from so small an object as a turnip plant, she cannot command the implement so effectually in the thinning of the plants as when the hands are placed nearer the working part of the hoe. This work is performed by the fieldworkers of the farm, and they are placed or *stented* to the work, as it is termed, at every two rows, that is, beginning at one side of a field, the first worker is stented to the first and second drills, the second worker to the third and fourth drills, the third worker to the fifth and sixth drills, and so on with every other worker. The reason for this particular arrangement, instead of giving one drill to every worker, is that each may have sufficient room to work, and, having two drills each, the whole body of workers have to shift their ground the seldomer. It is not easy to give a short account of the mode of using the hoe in singling turnips; but the following directions may serve to show the leading requisites to perform that operation in the best manner. On *commencing to single* the first drill, the feet are placed on each side of the second drill, so that the side of the worker is presented to the drill to be singled, the shaft of the hoe is held near its end with one hand, while the other hand, being that of the side in front, is placed a little in advance. The foremost hand indicates whether the person is right or left handed, as it is rare to find a worker that can single turnips with either hand. The foremost hand is steadied by being rested partially on the bend of the leg of the same side. The hoe is pushed chiefly by the weight of the body of the worker against the row of plants, when as many plants are removed by the forward push as the length of the face of the hoe covers, and the body is then brought back to its former position, so there is an oscillation of the body forwards and backwards in the act of singling. In pushing away another portion of the plants, the side of the hoe covers the plant next to the one intended to be left growing, which is, of course, made to stand single.

Improved hand hoe

This singling constitutes the difficulty of the operation for, if attention and dexterity are not exercised in it, the plant will be dragged up by the roots with the slightest hold of a portion of its leaf; and, although the leaves may be quite free, the stem or roots may intertwine with those of the adjoining plants. It is found that the best mode of avoiding these difficulties is to single the plants before the leaves become so enlarged as to be confounded with those of other plants, or the stems to become so drawn up as to intertwine amongst each other. It is also found that in pushing the hoe is a much safer mode of leaving the plants single than by drawing it towards the worker. The plant, on being left single, falls away from the worker upon its side, partly from want of the support of the others, and partly from the taking away of a portion of the soil from its roots. Singling of turnips should only be prosecuted when the ground is dry, and the plants themselves also dry, as they then separate from one another more readily. Whenever the ground becomes cloggy on the hoes, even with a shower, the work should be desisted from until drought returns.

HOEING

After the plants that have been removed by the singling into the middle of the drills are partly decayed, and the leaves of the single plants have attained about 6 inches [15cm] in length, the *horse hoe* or *scuffler* is again set to work to harrow the removed plants to the surface, and to destroy any surface weeds that may have made their appearance. After the turnips are all singled and scuffled, the fieldworkers hoe the turnips in the order they were singled, that is, remove the weeds from between the plants and loosen the soil immediately around them. This is done by setting one foot on each side of a drill and, on grasping the hoe short, the earth is loosened with it around every plant, all double plants removed by the hand, and every weed pulled by the hand that is found growing too near a single plant for the hoe to remove, which, if attempted by the hoe, might cut the plant through; and care should also be taken that no plants are cut through by the root underground with the hoe. A second hoeing of a similar kind generally finishes the hand culture of the turnip crop.

Swedes

Swedish turnips or swedes transplant very well, like the common cabbage; but the true turnip, the white globe or yellow, do not transplant, and any attempt to fill up blanks with them ends in disappointment.

Mangelwurzel

The culture of mangelwurzel is exactly similar to that of Swedish turnips, up to the point of sowing the seed, which is, however, done in a very different manner, owing to its peculiar structure; that of Swedish turnip is smooth, and can be sown with a machine, but the seed of mangelwurzel is contained within a persistent capsule which is so rough as to be impracticable for any machine yet invented to sow it. There are three kinds recommended for field culture, the common long red, the sugar beet and the orange globe beet. The way I sowed the common long red or marbled mangelwurzel, after the drills had been dunged and split in the double mode on good hazel loam, in the end of April, was this. A rut was made by a fieldworker along the top of the drill with a common draw hoe, and another fieldworker followed and dropped in the seed along the bottom of the rut with the hand, and a third followed last and covered up the seed with the earth raised by making the rut, with the back of an iron garden rake trailed along the rut. The crop was singled by hand, hand-hoed, and horse-hoed. The crop proved a good one; was taken up before the arrival of frost, stored amongst dry sand like carrots and given to milch cows in winter, who were very fond of it.

Rape

Rape is sown at two seasons, in autumn to be ready to be eaten as spring food, and in summer, to be consumed in autumn before the consumption of turnip commences. Its culture in summer is precisely that given to swedes, and as its seeds are smooth, though larger than turnip seed, they are sown by the same machine. The plants are not

thinned out in the drills like turnips, because it is the leaves, and not the root, which are used; but the ground is horse and hand-hoed until the leaves are so far advanced as to be able to keep down weeds. Rape is cultivated in this country solely for the use of sheep, for ewes before they are tupped in autumn, to bring them into season, and perhaps in preparation for turnips, and for great ewes and hoggs in spring, to bring the milk on the ewes. It is extensively cultivated on the continent, in Belgium, Holland and Germany, for its seed, out of which is expressed the rape oil that is much used in manufacture.

Carrots

Carrots are only occasionally raised in the field for farm use in this country, not but that the climate is quite congenial to their growth, but their growing best in light soils, the culture is necessarily confined to certain localities. Carrots will not succeed with dung applied directly under them, as they are then very apt to become much affected with worms, and to fork into a number of roots. The dung should be applied, and in large quantity, as for swedes, upon the soil in autumn.

The seed being confined in a capsule covered with small hooked spines, cannot be sown in a satisfactory manner with a machine, though attempts have been made to effect the purpose. It must therefore be

The village community come together to thresh the rape crop

sown with the hand, and as the capsules are apt to adhere together, they will separate more freely if mixed with a little dry earth, or still better with sand. A slight rut should be made along the top of the drills with a hoe, the seed strewed along and covered with earth. The plants should be singled out at the distance of 4 inches [10cm] by the hand, as their roots strike down too deep at once for the hoe to remove, without at the same time taking away too much of the soil from the top of the drill.

Parsnips

The parsnip requires a mild summer to be successfully raised in the field. Where it is desired to be raised, it may be cultivated as the carrot, either on drills, or in rows upon the flat ground, or broadcast, but it affects a much stronger soil than the carrot, though it may be raised upon sand, and even peat, if an additional quantity of manure is applied. The seed is contained in a broad, thin capsule, is very light, and 10lb [4.5kg] the acre are required when good. It should be sown in April. It should be new, and steeped before being sown, but if sown in a very dry soil, when soaked, it is apt to be destroyed by the drought. The thinning and weeding of the crop may be affected in the same manner as the carrot.

CLEARING STONES, REPAIRING FENCES, AND THE PROPER CONSTRUCTION OF FIELD GATES

The season being almost arrived when the grass is able to support stock and, of course, when the cattle are permitted to leave their winter quarters in the steading, it is necessary to ascertain, in the first place, whether the fences of the grass fields are in such a state of repair as will offer no temptation to stock to scramble through neglected gaps, much to the injury not only of the fence, but perhaps of themselves, or at least much to their disquietude; and, in the next place, to watch the period when the grass is in a fit state to receive them. Sometimes a good deal of work is required to put grass fields in a proper state for the reception of stock, owing principally to the nature of the soil, and partly to the state of the weather.

CLEARING STONES

On every kind of land the small stones lying on its surface should be gathered by the fieldworkers and carted off for the use of drains, or be broken into metal for roads. It may happen that the throng of other work may prevent the assistance of horses and carts being given for this purpose, in which case the stones should be gathered together in small heaps on the furrow brow of every other single ridge; but in doing this, it should be remembered that these heaps occupy so much of the ground and, of course, prevent the growth of so much grass, that, on this account, it is a much better practice to cart them away at once if practicable. When carts are used the stones are thrown directly into them; whereas in making heaps, the stones require some care to be put together, and, of course, waste time, and they have to be removed after all. Some farmers are regardless of gathering the stones from any of their fields, even from grass fields which are to be in pasture; while all acknowledge that fields of grass which are to be made into hay ought to be cleared of stones to save the scythes at hay time. On clay soils there are very few, or perhaps no stones to clear off, and in wet weather no cart should be allowed to go on new grass.

As every field, whether of new or of old grass, should be rolled some time before the stock enter them, it is clear that the ground cannot receive all the benefits of rolling as long as stones are allowed to remain on its surface. The best time for rolling is when the surface is dry – mark you, not when hard and dry – for when grass, especially young grass, is rolled in a wet state, it is very apt to become bruised and blackened. When dry, grass is elastic and able to bear the pressure of the roller without injury. Light land will bear rolling at any time when the surface is dry; but plants are very liable to be bruised by the roller against the hard clods of clay land and, in a soft state, on the other hand, clay land is apt to become hardened or encrusted by rolling. The rolling of heavy land is thus a ticklish matter; but a good criterion to judge of its being in a fit state for the roller, is when clods crumble down easily with the pressure of the foot, and not press flat, or enter whole into the soil. The rolling is always given across the ridges. The stones should be gathered, and the land rolled at least a fortnight before the stock are put on grass, to allow the grass time to grow after these operations, when it will be found to grow rapidly, if the weather is at all favourable.

Repairing the fences

While the surface of the field is thus preparing for the reception of stock, the hedger should be engaged in repairing the fences. In this he is frequently assisted by the shepherd and, in cases where there is no professed hedger on a farm, the shepherd himself undertakes the duty. The repairing chiefly consists in filling up gaps, and these are rendered fencible by driving stabs on the face of the hedge bank behind the gap, and nailing two or three short rails on them or by wattling them with branches of trees or thorn, or by setting a dead hedge (see page 144). There should nothing be put into the gap, as is often done, to the prevention of the lateral extension of the plants on either side of it, and which of themselves in time will fill up a narrow gap. A wide gap should be kept clear, and filled up in due season with living plants. Every gateway in a field not required for the season should be filled up with a dead hedge. Stone fences should be repaired by replacing the cope stones, and rebuilding the few stones that may have been driven down by violence.

In making repairs of fences, it should be borne in mind to keep an easy passage for the shepherd from field to field, when looking after his flock. Facilities should be afforded him, by leaving openings at the corners of fields, or setting stiles across the fence (see illustration); because it is better that these should be formed for him at once, than that he should have to make them for himself. He is the best judge of where they should be placed, in the short cuts he must necessarily take by the fields.

Fence steps

Gates

Besides the fences, the gates of grass fields require inspection and repairs, so as they may be put in a usable state for the season. When any of the timbers, posts or bars, are broken or wanting, or the fastenings loose, the carpenter or smith should be made to repair them; and the posts on which the gates hang should be made firm in the ground when loose, or renewed when decayed. In putting up new gate posts, the firmest mode I have found is to dig as narrow a hole as practicable 3 feet [91cm] deep for the hanging post, and then to ram the earth, by little and little, firmly around the post without any stones. Charring or pitching the part underground is a pretty good prevention from rot for some time. The simplest mode I have seen of fastening field gates is with a small chain attached to the fore stile of the gate, to link on to a hook on the receiving post. The most convenient position for field gates is at the ends of head ridges, which may be regarded as the roads of fields. Field gates should be made to fold back upon a fence; to open beyond the square; and not to shut of themselves. When they shut of themselves, and are not properly set when opened, and which requires greater care than is usually bestowed on these matters, they are apt to catch a wheel of the cart which is passing and, of course, to be shivered to atoms, or the post to be snapped asunder; and more than this, self-shutting gates are apt to be left unfastened by most people who pass through them, and are therefore unavailing as a fence to stock, especially to horses when idle, which seem to take delight to loiter about gates, and they not infrequently find out the mode of opening them.

Secure mode of fixing the hanging post of a field gate

A gate, generally speaking, may be described as a rectangular frame; there are exceptions to this definition applicable to gates as a whole, but to field gates there are none. A gate, to be permanent, should be immutable or unchangeable in its form, a simple rectangular frame without upfillings, or even with upfillings, if they are placed at right angles to each other, is the most liable to change of any connected structure of framework.

The triangle, on the other hand, is the most immutable or least liable to change; it is, in short, so long as the materials remain unchanged, perfectly immutable, but a gate in the form of a triangle would, in most cases, be very unserviceable, though a combination of triangles may produce the requisite figure for a serviceable gate. If then we take the rectangular frame so essential to a field gate, and apply a bar in the position of the diagonal of the parallelogram, we immediately convert the original rectangular figure into two triangles, applied to each other by their hypotenuse, and this gives us the true elements of a properly constructed gate, all the other parts being subordinate to these, and adapted solely to the practical purposes of the gate as a defence or for ornament. In many cases, depending upon the material employed, an opposite diagonal may be applied, dividing the gate into four triangles; but, in general, this is only necessary where flexible rods of iron are applied as the diagonals.

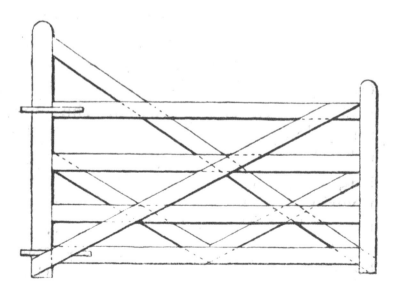

Strong field gate

WEANING CALVES AND BULLS, AND GRAZING CATTLE TILL WINTER

The cattle which were accommodated in their respective places in the steading at the beginning of winter continue to be treated through the spring months in the same manner. In fact, their treatment is throughout the same until turned out to grass, which is usually some time in May; unless variety of food may be regarded as a difference. It is found that cattle in a state of confinement in a steading thrive better on a variety than on the same food; and yet when on grass they require no variety of food, and thrive the better the longer they are kept upon it.

The grass should be ready to afford a bite for cattle whenever the turnips are all exhausted; at which time the cattle will be found to be in this state: the two-year olds, now three-year-olds, will be full fat, and ready to be disposed of to the butcher or dealer; the one-year-olds and calves will have grown much to bone, and their condition will have improved in proportion to the quantity of nourishing food they have received during the feeding season; the cows will all have calved, or should all have calved, for a May calf is too late to bring up and form a part of the herd – they will be in milk and in fresh condition to put to grass; the early calves will have just been weaned, and in excellent order to put on grass, and the remainder will be in rapid progress towards weaning. These cattle, in their respective states, are treated in summer in a different manner from each other, and different from the way they were treated in winter, and they are treated thus.

The fat cattle are seldom put on grass, being disposed of to the butcher or dealer, either at home or in markets held for the purpose.

Ayrshire cow

Breeders of their own stock seldom dispose of their fat cattle until the turnips are nearly consumed, being anxious to keep them as long as possible, for the sake of the excellent manure which the turnips afford. Those who purchase oxen to fatten usually buy a larger lot than can be maintained on full turnips till the grass is ready, in order to dispose of them before the season arrives when fat oxen are usually sold. Such fatteners of cattle dispose of their lots in whole, or in part, from the end of the year to April, whenever the market offers them the most profit.

Pasture for cattle

The fat cattle being disposed of, the pasture should be judiciously distributed amongst the remaining, the object of keeping them being to breed calves and to afford plenty of milk to bring them up; the more milk they yield the better will the calves prove, and the more profitable they will themselves prove after the calves are weaned.

It should be kept in mind that cows are peculiarly susceptible to sudden changes of temperature, especially from heat to cold and from drought to rain, so that whenever cold or rain, or both together, which is the most common circumstance, occur, they should be brought into the byre. For some time after they are put out to grass they should be brought into the byre at night, where they are milked, and again in the morning before they are let out to the field, and milked in the field at midday. After the nights become warm, I have found it conducive to health, and it is both a rational and a natural plan, to allow them to lie out in the field all night, and to milk them there at stated hours, three times every day, the shepherd or cattleman taking it as a part of his duty to bring them to a certain spot of the field to be milked, and which is usually named the *milking loan*. This mode of allowing them to lie out always in a sheltered field no doubt imposes a good deal of labour on the dairymaid and her assistant in carrying the milk to the dairy after the calves have been weaned, but I am persuaded it is an excellent system for the health of the cows.

Under it, cows rise from their lair at daybreak and feed while the dew is still on the grass, and by the time of milking arrives, say 6 o'clock, they are already partially filled with food, and stand contented, chewing the cud, while the milking proceeds. They then roam and fill themselves and, by 9 o'clock, lie down in a shady part of the field and chew their cud until milking time arrives at midday, when they are again brought to the loan and milked. Again they roam for food, and, when the afternoon is hot, will stand in the coolest part of the field whisking away the flies with their tail and ears. The evening milking takes place about 7pm, after which they feed industriously, and take up their lair about sunset, and from which they rouse themselves in the morning before being milked.

The lying out at night, too, saves the trouble of providing supper for the cows, which they must have when housed in the byre. But whenever the weather becomes cold and wet, cows should be brought into the byre at night, and supplied with supper, such as cut aftermath [second mowing or crop of hay in that season] or tares.

Weaning

The weaning of calves should not exceed one month after the cows have been on grass, that is, by the end of June, for a calf later weaned than that period has been too late brought into the world to be worthy of belonging to the standing stock of a farm. As cows increase in gift of milk after the grass has fairly passed through them, the late calves should have as large an allowance of new milk, three times a day, as the quantity obtained will allow, reserving a little for the use of the house. The most convenient first grass field for calves is a contiguous paddock, from which they should be brought into the court for a few nights and receive turnips and hay, until the grass has safely passed through them, and the weather prove mild and dry for them to lie out all night in the paddock. The youngest calves should now leave their cribs and pass a few days in the court until they become accustomed to the air and sun, when they also may be put into the paddock during the day, and there supplied with their diets of milk, and brought into the same court at night until they are able to be out all night. In weaning the youngest calves, the milk should be gradually taken from them until they take with the grass, upon which they must then entirely depend. A little after all this has happened, say, by the middle of July, the pasture in the paddock will become rather bare, and the whole lot of calves should then be taken to good pasture, where they will have a full bite, for nothing can be more injurious for calves than to place them on bare pasture to fall away in condition immediately after weaning, and which they will assuredly rapidly do, and from which it will be very difficult to recover them all summer.

Bull calves

Bull calves should have good milk every day until the grass is able to support them, in order to strengthen their bone and maintain their condition. When a number are brought up together, they should be grazed by themselves on the best grass the farm affords, or they may go along with the cows. The young one-year-old bulls should now be furnished with a ring in their nose. This instrument is useful not only in leading the animal, but, being constantly in use, in keeping his temper in subjection. I have no doubt whatever that such a ring affords the most complete command over the most furious bull. In case of a bull becoming more irritable and troublesome as he advances in years, which many bulls are inclined to be, the ring furnishes the means of curbing him at once, when it would otherwise be impossible to get hold of his nose. It affords also an easy means of suspending a light chain from the nose to the ground, upon which the forefeet are ready to catch the chain in walking, when the nose receives so sudden a check that whenever the bull attempts to run at anyone in the field, he pains himself. Even a young bull in a field may follow you at first in sport, and run at you afterwards in earnest.

The ring is put into a bull's nose in this way. Let a ring of iron be provided, of perhaps 2½ inches [6.5cm] in diameter overall. It should have a joint in it, to let the ring open wide enough to pass one end through the nose, and the two sides of the ring, on being closed again after the operation, are kept together with two countersunk screws (see the illustration right). An iron rod tapering to the point, and stouter than the rod of the ring, should be provided. Let a cart rope have a noose cast firm at its middle, and put the noose over the bull's head, and slip it down his neck, with the knot undermost, till it rests upon the breast. Any mortared wall sufficiently low to allow the bull's head to reach over it will answer to put him against; or what is safer for his knees, any gateway with a stout bar of wood placed across it as high as his breast. Place the bull's breast against the wall or bar, and pass the rope from the lowest part of the neck along each side round the buttock like a breeching, and bring one end of the rope over the wall or bar on each side of the bull's head, where a stout man holds on at each end, and it is the duty of both these men to prevent the bull from retreating backwards from the wall or bar. A man also stands on each side of the bull's buttock to prevent him shifting his position. The operator having the iron rod given him heated in the fire, just red enough to see the heated part in daylight, he takes the bull by the nose with his left hand and, feeling inwards with his fingers, past all the soft part of the nostrils, until he reaches the cartilage or septum of the nose, he keeps open the nostrils, so as on passing the hot iron through the septum, it may pass clear through without touching the outer skin of the nostrils, taking care to pass the iron parallel to the front skin of the nose, otherwise the hole will be oblique. Immediately after the rod has been passed so far as to make the hole sufficiently large, and the wound has been sufficiently seared, the operator takes the ring, opened and, still keeping hold of the bull's nose with the left hand, passes it through the hole and, on bringing the two ends together, puts in the screws and

Opened bull's ring

Closed bull's ring

secures them firmly with a screwdriver. On being satisfied that the ring turns easily round in the hole, and hangs or projects evenly, the bull is then released.

The ring should not be used until the wound of the nose is completely healed. The readiest and neatest way to attach a rope to a bull's ring is with a swivelled hook, retained in its place by a spring, and a rope should be kept for the purpose. On first trying to lead a bull by the ring, the drover should not endeavour to pull the animal along after himself, but allow him to step on while he walks by his side, or even behind him, with the rope in his hand. While so following, to relieve the animal as much as practicable of the weight of the rope upon the nose, the drover should throw the middle of the rope upon the bull's back, and retain a hold of its end. Should he offer to step backwards, a tap on the shank with a stick will prevent him; and should he attempt to run forward, a momentary check of the rope will slacken his pace. On no account should the drover attempt to struggle with the bull on the first occasion; on the contrary, he should soothe and pacify him, and endeavour to inspire him with confidence in himself and the rope, and to show him that he will receive no hurt if he will but walk quietly along. The animal, in the circumstances, will soon learn the nature of the tuition he is undergoing if he is properly dealt with, but if tormented merely that the drover may show his power over a powerful animal, it may be a long time, if ever, before he will learn to behave quietly when led.

A useful instrument for leading a bull by occasionally, when he has not been ringed, or for leading a cow to the bull at some distance, or for taking away any single beast, and at the same time retaining a power over it, is what is named the *bullock holder*. It consists of iron in two parts jointed, which are brought together or separated by a thumbscrew passing through them. The ends furthest from the joint terminate in a ring having an opening at its extreme side, each end of which opening is protected by a small ball. The arms of this ring embrace the septum of the nose gently between them; and the shank of the instrument being screwed close together, the balls approach no closer than just to embrace the septum, and the nose of the animal prevents them slipping out. The leading rope is attached to the jointed end of the instrument, which is formed for the purpose into a small ring.

Scoop for filling the water barrel

because in cold and rainy weather it is scarcely visited by the cattle, in hot weather it is supposed to be viewed with the same indifference. And, even where tanks are duly attended to for cattle, there are none set down at a lower level for sheep. A watering pool should be securely fenced, as cattle are very apt to push one another about while in it and, for that reason, it should also be roomy. It should be of considerable length and narrow, to allow access to a number of animals at the same time, if they choose to avail themselves of it; and I have often observed cattle delight to go to the water in company. Pools are usually made too small and too confined. The access to them should be made firm with broken stones in lieu of earth, and gravel placed on its bottom keeps the water clean and sweet, while the water should flow gently through the pool.

Shade for cattle

The want of shade in pasture fields is also a sad reflection on our farmers. Observe, in summer, where the shade of a tree casts itself over the grass, how gratefully cattle resort to it, and where a spreading tree grows in a pasture field, its stem is sure to be surrounded by cattle. The stirring breeze under such a tree is highly grateful to these creatures; and such a place affords them an excellent refuge from the attacks of flies. In cold weather, also, observe how much shelter is afforded to cattle by a single tree, and how they will crowd to the most wooded corner of a field in a rainy day even in summer. Ought not such indications of animals teach us to afford them the treatment most congenial to their feelings? I am no advocate for hedgerow trees, even though they should cast a grateful shade into a pasture field, and still less do I admire an umbrageous plane in the middle of a field that is occupied in course with a crop of grain or turnips; but similar effects as good as theirs may be obtained from different agencies. A shed, erected at a suitable part in the line of the fence of a field, would not only afford shade in the brightest day in summer, but comfortable shelter on a rainy day, or on a cold night in autumn. Such an erection would cost little where stone and wood are plentiful on an estate, and they could be erected in places to answer the purpose of a field on either side of the fence when it was in grass.

Cattle drink from a stream which cools the hooves to prevent foot sore at the height of summer

SHEEP WASHING, SHEEP SHEARING AND WEANING LAMBS

I have said that, as lambs become strong enough to be put to pasture, they always get new grass in order to increase the milk of the ewes. The new grass, to be pastured by ewes and lambs, should be selected with judgment, and that intended for hay should first be stocked, because it is found that new grass, if moderately eaten down in spring, stools out, and affords a thicker cutting at hay time, than if it had not been so pastured. For the same reason, the new grass intended to cut for horses' forage, should also be earlier pastured than what is to be

A team of shearers set about shearing the flock

pastured all the season, not only to give both it and the hay grass time to attain their growth when they shall be wanted, but to give the pasture grass time to become so strong as to support being pastured. None of the new grass should be eaten too bare, even the part which is to be pastured by the ewes; and rather than commit such a mistake, even in a late season, the ewes should have a hasty run over the best of the older grass for a fortnight or so, till the hained new grass has revived. After the castration of the lambs there is nothing to do to them until the ewes are washed preparatory to shearing the wool from them, and which is done about the beginning of June.

Washing

The season for washing sheep having arrived, a fit place should be selected for the purpose. It should consist of a natural rivulet or, where that is wanting, of a large ditch, having both its banks clad with clean sward. The next step is to form a damming across the rivulet, if it is not naturally sufficiently deep of water to conduct the operation of washing. The bottom of the river or ditch should be hard and gravelly, and the water in it pure, or it will not answer the purpose, as a soft and muddy bottom, and dirty water, will injure the wool more than do it good.

A *damming* may be made either entirely of turf wall built across the stream, though that imposes considerable labour and waste of grass, or with an old door or two or other boarding placed across the bed of the stream, supported by stabs against the weight of water, and the chinks at the bottom and sides filled up with turf; and over which, when the water accumulates, the water falls. In constructing this dam, the overflowing should be as great as to cause such a current in the pool as to carry away quickly all impurities. One side of the pool is occupied by the unwashed, and the opposite by the washed sheep. They are confined in their respective places by flakes or nets. To prevent any sheep taking the water of themselves, which they are apt to do when they see others in before them, the fence should be returned along the sides of the pool as far as the men who wash the sheep are stationed (see the illustration on page 215). The damming *a a*, by means of doors and stakes, and turfing, retain the water until it overflows. The net placed on each side of the pool is returned so far down both its sides. The depth of the water is seen to take the men to the haunches – the proper depth. Everything being thus prepared at the pool, the sheep are also prepared for the washing. The lambs not being washed, and to save trouble with them at the washing pool, they are separated from their mothers, and left in a court of the steading until the washing is over. The ewes, hoggs [young sheep] and dinmonts [one-year-olds] are all taken to the pool in a lot. They should be driven gently, and not allowed to be at all heated. The ewes will be troublesome to drive, being always in search of their lambs; but notwithstanding this annoyance, they should not be dogged, but rather give them plenty of time upon the road. They should be driven along the road most free of dust or mud. The men who are to wash also prepare themselves by casting their coats, rolling up the sleeves of their shirts, and putting on old trousers and shoes to stand in the water with. The shepherd and other two men are quite enough to wash a large number of sheep thoroughly but, if the stream is very broad, another may be required to save time in handing the sheep across. These three men are represented in the figure, e being the shepherd, and the last man to handle the sheep, and d and c his assistants. At least two men are required to catch the sheep for the washers, of whom one is seen at b. On an occasion of this kind the men receive a gratuity of bread and cheese, and ale, and also a dram of spirits as a stimulus, as at h, where the dog is seen to keep watch.

The washing is performed in this way. While the three men are taking up their positions in the water, the other two are catching a sheep; and to render this fatiguing work more easy, the fold should not be made larger than to contain the sheep easily. A sheep is caught, and is being presented by b to the first washer c, who takes the sheep into the water and, allowing its wool to become saturated with it, turns it over upon its back, keeping up its head, and taking a hold of the arm of a foreleg with either hand, and of the wool on the opposite side of the head with the other. He then dips the sheep up and down, to and fro, and from one of its sides to the other, slowly, causing all the wool to wave backwards and forwards, as if rubbing it against the water. In doing this the water becomes very turbid about the sheep, and he continues to do it till it clears itself, when he hands the sheep to the next washer d, standing in the middle of the stream. Whenever c gets

Washing pool and sheep washing

quit of one sheep, another should be ready by the catchers for him to receive into the water. The second washer *d* holds and manages the sheep in the same manner; and then hands it to the shepherd *e*, and is immediately ready to take another sheep from the first man. It is the duty of the shepherd to feel if the skin of the sheep is clean, and every impurity removed from the wool. The position of the sheep on its back is favourable for the rapid descent of earthy matter from the wool. Wherever he feels a roughness upon the skin, he washes it off with his hand, and wherever any clots are felt in the wool which have escaped the other washers, he rubs them out. The belly, breast and round the head he scrubs with the hand. Being satisfied that the sheep is clean, he dips it over the head while turning it to its natural position, when it swims ashore and gains the bank at *g*. Its first attempts at walking on coming out of the water are very feeble, its legs staggering under the weight of the dripping fleece; in a little after it frees itself from the water entirely by making its fleece whirl like a large mop. After washing, sheep should not be driven along a dirty or dusty road, nor should they be put into any grass field with bare earthy banks, against which they might rub themselves. In fact, they should be kept perfectly clean until their fleeces are taken off.

Shearing

IN PREPARATION FOR SHEARING A place under cover should be selected for clipping the fleeces. The clipping floor is prepared in this way. Let clean wheat straw be spread equally over the floor 2 or 3 inches [5 or 7.5cm] thick, and spread over it the large canvas barn sheet, the edges of which should be nailed down tight to the floor. The use of the straw is to convert the floor into a sort of soft cushion for the knees of the clippers, as well as for ease to the sheep. A broom should be provided to sweep the cloth clean. A board 2½ feet [75cm] above the floor should be provided to wind the fleeces upon, and it may be placed along the wall of the barn opposite to that occupied by the clippers. A space near this should be cleaned, and a sheet spread upon it, for putting the rolled fleeces upon. The remainder of the barn should be cleared of dust both from the floor and walls as high as the sheep can reach, and a little clean straw strewn upon the floor for them to lie upon.

The instrument by which the wool is clipped off sheep is named *wool shears* (illustrated below). They require no particular description farther than to explain that the bend or bowl a, which connects the two blades, acts as a spring to keep them separate, while the pressure of the hand on each side of the handle b overcomes and brings the blades together.

In case of dew or rain in the morning, it is customary to bring into the barn as many dry sheep on the previous evening as the clippers will shear on the ensuing day.

It is a frequent custom for neighbouring shepherds to assist each other; and though the plan does not perhaps expedite the entire sheep shearing of the country, yet a number of men clipping at the same time makes work seem lighter, and it gets the clipping of any individual flock the sooner through. Clipping being both a dirty and heating work, the coat should be stripped, and the oldest clothes worn; and the hat and the vest are commonly thrown aside. Garters or tight knee breeches will be found very irksome pieces of dress in clipping.

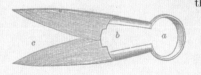

Wool shears

FIRST STAGE Clipping is done in this way. Whenever a sheep is caught in the barn, every straw or dirt on the wool or hoofs should be taken away before it is laid on the canvas carpeting. Clipping consists of three stages, the first of which is represented in the illustration. After setting the sheep on its rump and, on the supposition that the clipper is a right-handed man, he goes down on his right knee and leans the back of the sheep against his left leg *a*. Taking the shears in his right hand and holding up the sheep's face with his left, he first clips the short wool on the neck, and passes down the throat and breast between the forelegs to the belly. Then, placing the forelegs under his left arm, as seen at *b* under *c*, the belly is left exposed to be next shorn across from side to side down to the groins. In passing down here, while the shears *d* are at work, the left hand *e* is engaged keeping the skin tight where it is naturally loose. The scrotum *f* is then bared, then the inside of the thighs *g g*, and then the underside of the tail *h*. These complete all the parts of this position.

The shears are used in a particular manner, to be safe alike to the fleece and the skin of the animal. The essential thing is to keep the points always clear of the skin, for if held downwards, they will inevitably run into it, and should such a prick be made when the hand is about to close, the consequence will be that a large piece of the skin will almost be clipped out before the clipper is aware of what he is about. This is a common error committed by new clippers, and it is a great offence in any clipper's hands. The only way to avoid this serious injury to sheep is to rest the broad part of the shears only and always upon the skin. In this position, with the skin drawn tight by the left hand, the shears are made to move forwards with a hold of the wool not exceeding 1 inch [2.5cm] in breadth, in very short and frequent clips, taking care only to bring together the broad parts of the blades from where they are seen to separate to as far as *c* (see left), keeping the points always apart.

First stage of clipping a sheep

SECOND STAGE The second stage of clipping is a position gained by first relieving the forelegs from the first position, and gently turning the sheep on its far side, the forelegs *c* are put under the right or clipping arm *f*, while the clipper, going on both knees, supports the shoulder of the sheep upon them, thus giving the animal an easy reclining posture (shown overleaf). You may rely upon this fact that the more at ease the

Second stage of clipping a sheep

animal feels, the more readily will it lie quiet to be clipped. Supporting the head of the sheep with his left hand, the clipper first removes the wool from behind the head, then around the entire back of the neck to the shoulder top. He then slips its head under his left arm, as *a* under *g*. Having the left hand thus at liberty, he keeps the skin tight with it while he clips the wool with the right, from where the clipping in the first position was left off to the backbone. In the figure the fleece appears to have been removed about halfway down the carcass, the left hand *b* is laid flat, keeping the skin tight, while the right hand *e* holds the shears at the right part and in the proper position. The clipper thus proceeds along the thigh and the rump to the tail *d*, which is entirely bared at this time.

THIRD STAGE Clearing the cloth of the loose parts of the fleece, the clipper, holding by the head, lays over the sheep on its clipped side and, still continuing on his knees, slips his left knee *a*, over its neck to the ground, while his left foot *b*, resting on the toe, supports the left leg *c* over the neck of the sheep, and keeps its head *d* down on the ground. This is the third position in clipping (below left). The wool having been bared to the shoulder in the second position, the clipper has now nothing to do but to commence where it was left off in the first position and clear the fleece entirely to the backbone, meeting the clips where they were left off in the second position, the left hand *e* being still at liberty to keep the skin tight, while the right hand *f* uses the shears along the whole side to the tail. The fleece *g* is now quite freed from the sheep. In allowing the sheep to rise, care should be taken that its feet do not become entangled in the fleece, for in its eagerness to escape from the unusual treatment it has just received, its feet will tear the fleece to pieces. Immediately that the lot of sheep in the barn is clipped, it is taken to the field, and another brought in its place.

Third stage of clipping a sheep

A NEW-CLIPPED SHEEP A new-clipped sheep should have an appearance where the shear marks are seen to run in parallel bands round the carcass, from the neck and counter, along the ribs, to the rump and down the hind leg. When pains are taken to round the shear marks on the back of the neck; to fill up the space in the change of the rings between the counter and the body; to bring the marks down to the shape of the leg; and to make them run straight down the tail, a sheep in good condition so clipped forms a beautiful object. A sheep clipped to perfection should have no marks at all, for they are formed of small ridglets of wool left between each course of the shears; but such nicety in clipping with shears is scarcely possible and, at any rate, the time occupied in doing it would be of more value than all the wool that would be gained. It should be borne in mind, however, that the closer a sheep is clipped, it is the better clipped, and is in a better state for the growth of the next year's fleece.

WEANING LAMBS

Clipping makes such a change on the appearance of sheep that many lambs have difficulty at first in recognizing their mothers, whilst a few forget them altogether, and wean themselves, however desirous their mothers may be to suckle them; but as the ewe is content with one lamb, many a twin which does not follow her is weaned on this occasion.

Leicester lambs are weaned at the end of June or beginning of July; and the process is simple and safe, as most of them by that time chiefly depend upon grass for support. All that is required is to separate the lambs from the ewes, in fields so far lying asunder, as to be beyond the hearing of the bleatings of each other. Where there is the convenience, lambs should be put on hilly pasture for some weeks at this time, the astringent quality of which gives an excellent tone to their system, and renders them more hardy for winter. Some farmers even hire rough hill pasture for their lambs, but where such cannot be had, they are put on the oldest, though good pasture, for a few weeks before the aftermath [second crop of hay] is ready to receive them.

The ewes, when separated from their lambs, should be kept in a field of rather bare pasture, near at hand, until their milk be dried up. They must be milked by the hand, for a few times, till the secretion ceases – once, 24 hours after the lambs are taken away, again, 36 hours thereafter, and the third time perhaps two days after that. Even beyond that time a few may feel distressed by milk, which the shepherd should relieve at intervals until the udders become dry. Indeed, milking after weaning of lambs is essential to the safety of ewes, and I fear it is not so effectually performed as it should be until the udders go dry.

ROLLING THE FLEECE, AND THE QUALITIES OF WOOL

Whenever a fleece is clipped from the sheep, a field worker should be ready to roll it up. I have already said that a board should be provided to roll the fleeces upon. Suppose the board to be placed in the clipping barn, the fleece, whenever separated from the sheep, is lifted unbroken from the cloth, and spread upon the board on its clipped side.

The winder examines the fleece that it is quite free of extraneous substances, such as straws, bits of thorn, of whin or burs, and removes them, and she also removes, by pulling off, all locks that have lumps of dung adhering to them, and which have escaped the notice of the washers. The winder being satisfied that there are no impurities in the fleece, folds in both its sides, putting any loose locks of wool into the middle, and making the breadth of the folded fleece from about 24 to 30 inches [61–76cm], according to its size. She then begins to roll the fleece from the tail towards the neck as tightly and neatly as she can and, when arrived at the neck, draws the wool there as far out, twisting it into a sort of rope, as will go round the fleece, hold its own end firm, and make the entire fleece a tight bundle (shown left). The fleece is then easily carried about, having the clipped surface outside which, being composed of wool saturated with yolk [natural oil], exhibits a shining silvery lustre.

Rolling a fleece of wool

Assessing good wool

Good wool should have the following properties: the fibre of the staple – a staple being any lock that naturally sheds itself from the rest – should be of uniform thickness from root to point, it should be true, as the phrase has it; the finer the wool, the smaller is the diameter of the fibre; the fibre should be elastic, and not easily broken; its surface should have a shining silvery lustre; and it should be of great density or specific gravity.

Of a staple all the fibres should be of the same length, otherwise the staple will have a pointed character; the end of the staple should be as bright as its bottom, and not seem as if composed of dead wool; the entire staple should be strong, and its strength is tested in the following manner. Take the bottom of the staple between the finger and thumb of the left hand, and its top between those of the right and, on holding the wool tight between the hands, make the third finger of the right hand play firmly upon the fibres, as if in staccato on the strings of a violin, and, if the sound produced be firm and sharp, and somewhat musical, the wool is sound; if the fibres do not break on repeatedly jerking the hands asunder with considerable force, the staple is sound; if they break, the wool is unsound and, what is remarkable, it will break at those places which issued from the felt of the sheep when the sheep was stinted of meat or had an ailment; though it will not break at every place simultaneously, because the weaker part, occasioned by the greater illness, will first give way.

A good fleece should have the points of all its staples of equal length, otherwise it will be a pointy one; the staples should be set close together; and it should be clean. One essential good property of wool is softness to the feel like silk, which does not depend on fineness of fibre, but on a peculiar property of yielding to the touch at once, and readily returning to the hand. There should be no hairs in wool, neither long ones, which are easily distinguishable from wool, and give the name of *bearded* to the fleece; nor short ones, soft and fine, like cat's hair, which are not easily distinguishable from wool, and are denominated *kemps*. The long hairs are frequently of a different colour from the wool; but the kemp hairs are of the same colour and, of the two, the latter are much the more objectionable, as being less easily detected.

Weighing the fleeces and packing the wool sacks

Making Butter and Cheese

The opening of the Cheese Market, Chippenham, Wiltshire, 1850

The dairy operations on a farm of mixed husbandry are limited, both in regard to the season in which, and the quantity of materials by which, they can be prosecuted. Until the calves are all weaned, which can scarcely be before the end of June, there is no milk to spare to make butter or cheese, but what of the former may suffice for the inmates of the farmhouse; and as some of the cows, at least, will have calved four months before all the cows are free to yield milk for the dairy, a full yield of milk cannot be expected from them even when entirely supported on grass. But though thus limited, both in regard to length of time and amount of milk, there is ample opportunity for performing every dairy operation, according to the taste and skill of the dairymaid. For example, butter may be made from cream, or from the entire sweet milk. It may be made up fresh for market, or salted in kits for families or dealers. Cheese may also be made from sweet or skimmed milk, for the market; and any variety of fancy cheese may be made at a time; such as cream cheese, imitation Stilton, Gloucester or Wiltshire.

Utensils

White Wedgwood-ware milk dish

The utensils with which a dairy should be supplied comprise a large number of articles, though all of simple construction. The *milk dishes* are composed of stoneware, wood, metal or stone. The stoneware consists of Wedgwood and common ware; the wooden of cooper work, of oak staves bound with hoops of iron; the metal of block tin or of zinc; and the stone are hewn out of the block and polished. Besides these, utensils formed of a combination of materials are used, such as wooden vessels lined with block tin or zinc, and German cast-iron dishes lined with

FESTIVITIES AT CHIPPENHAM.—THE HIGH-STREET, FROM THE BRIDGE.—ARRIVAL OF MR. NEELD, M.P.

FESTIVITIES AT CHIPPENHAM.

OPENING OF THE CHEESE-MARKET.

THE town of Chippenham, the scene of the festivities illustrated in the accompanying Engraving, is situate on the Great Western Railway, ninety-three miles from London, and thirteen from Bath, and forms the junction with the Wilts and Somerset Railway. The name is derived from its market, for which it was long known. It is an important seat for the manufacture of West of England broad cloths; and is fast acquiring celebrity from its great monthly markets for the sale of cheese, corn, cattle, &c., which have been established about sixteen years. For these benefits the town is principally indebted to one of its representatives, Joseph Neeld, Esq., a gentleman well known for his munificent patronage of the arts; and whose seat at Grittleton, Wilts, is enriched with some of the finest works in the kingdom.

Among other boons which Mr. Neeld has conferred on the inhabitants, is the erection of a new town or Market Hall, with the addition of an extensive market yard and sheds for the pitching of cheese, corn, &c. The hall is situated in the High-street, and was built some fifteen years ago by Mr. Neeld, at an expense of about £12,000, and placed by him at the disposal of the town. It has been since extended by the same gentleman three times, at an additional outlay of from £4000 to £5000. The alterations and extensions, the completion of which the demonstrations of this day were intended to celebrate, consist of the erection of an entire new shed of most commodious dimensions, for the cheese market, a very convenient exchange-room, and additional accommodation in the yard. The Cheese-market has a substantial and tasteful freestone frontage to the High-street, surmounted by the old borough arms in carved stone, beneath which are the words "Unity and Loyalty." The hall itself is 50 feet in length by 33 in width, and 19 in height, with spacious ante-rooms. The whole has been designed and executed by Mr. Thomson, of London, the architect of the Polytechnic Institution; and who has also, we are informed, erected several entire villages and churches, with school-houses, for Mr Neeld on his extensive estates.

Chippenham is surrounded by some of the best pasture land in the county, and, with its facilities of railway communication, its market has rapidly grown in importance; and, to meet this increase, Mr. Neeld has provided additional room for the pitching of cheese, so that the whole area now covered in for that purpose contains 15,500 superficial feet. A handsome room, named the "Exchange," has also been added.

The inhabitants of the town, to testify their sense of the important public benefits conferred upon them by Mr. Neeld, took advantage of the opportunity which the re-opening of the market yard afforded, to invite that gentleman to a public banquet on the 12th instant, and the demonstration then made was very striking. The whole of that part of the town through which the procession was expected to pass presented the appearance of rejoicings for some great triumph. Fir trees, planted for the occasion, lined the streets and road; triumphal arches were erected at almost every available point; festoons of evergreens and flowers extended across the streets in thick succession, whilst almost every house was profusely ornamented with laurels, flowers, and mottoes of welcome: it was a universal jubilee. The church bells rang, cannons fired, business was suspended the whole town was an emblazonment of flags and banners; every man and every woman and child waved welcome to Joseph Neeld.

At two o'clock, the Mayor and Corporation assembled at the Market Hall, and thence proceeded with the inhabitants towards Grittleton. An extensive procession was formed along the Grittleton road to the north of the viaduct, where it met Mr. Neeld, accompanied by a large body of his tenants, who had met at a luncheon provided for them at Grittleton House. After loud and hearty greeting, the cavalcade turned towards the town, in the following order of procession:—

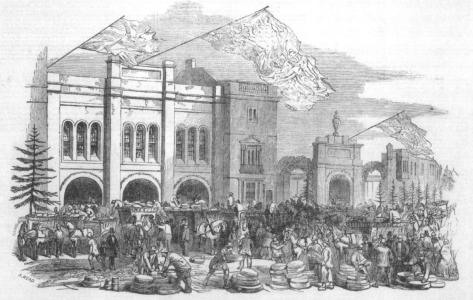

OPENING OF THE GREAT CHEESE-MARKET, AT CHIPPENHAM, SEPTEMBER 12, 1850.—THE MARKET HALL.

porcelain. Of the whole variety, the stone and wooden ones lined with metal are stationary, and the rest movable. All milk dishes should be of a broad and shallow form, for the purpose of exposing a large surface with a shallow depth of milk, in order to facilitate the disengagement of the several parts of the milk.

The other utensils are: *creaming scallop*, for taking the cream off milk; a *jar* for containing the cream until it is churned (a Wedgwood one, with the top and opening in it to be covered with muslin, to keep out dust and let in air, costs from 6s. to 7s. 6d.]; a *churn*, of which there are many forms, a *flat wooden kit*, to wash butter in; *scales* and *weights* for weighing butter, whether in pounds, fractional parts of a pound, or in the lump; *jars* or *firkins* [small wooden barrels] for packing salted butter; *moulds* for forming prints of butter for the table; *covered dishes* for holding fresh butter in pounds; a *tub* for earning the milk in, when about to make cheese; a *curd cutter* and a *curd breaker*; a *drainer* to lay across the cheese tub while the whey is straining from the curd; *cheese vats* for giving the form to cheese; a *cheese press*, a *furnace* and *pot* for heating water and milk; and a supply of *spring water* is an essential concomitant to a dairy.

CLEANLINESS

A word or two also on cleanliness. Unless the milkhouse is kept thoroughly clean, in its walls, floors and shelves, the milk will become tainted. In order to keep them clean, the floor and shelves should be of materials that will bear cleansing easily and quickly. In most farmhouses the shelving is of wood and the floor of pavement or brick. Wooden shelves can be kept clean, but are too warm in summer. Stone shelving is better, but must be polished, otherwise cannot be sufficiently cleaned; and to be kept clean, requires at times to be rubbed with sandstone. Marble shelving is the best of all for coolness and cleanliness, and they are not so expensive as many imagine. Polished pavement makes a more durable, easier cleaned and cooler floor than brick. There should be ample means of ventilation in the dairy when required; the principal object, however, not being so much a constant change or a larger quantity of air, as an equality of temperature through summer and winter. To obtain this desideratum, the windows, which should face the north or east, should not be opened when the temperature of the air is above or below the proper one, which, on an average, may be stated at 50°F [10°C]. The milkhouse should be thoroughly dry – the least natural damp in the walls and floor will emanate a heavy fungus-like odour, very detrimental to the flavour of milk and its products. The utensils should all be kept thoroughly clean and exposed to and dried in the open air.

However effectual woollen scrubbers may be in removing greasiness left by milk and butter on wooden articles, they should never be employed in a dairy, but only coarse linens, which should always be washed clean in hot water without soap and dried in the air. All the vessels should be quickly dried with linen cloths, that no feeling of clamminess be left on them, and then exposed to the air. In washing stoneware dishes, they should not be dried at that time, but set past singly to drip and dry; and they should be rubbed bright with a linen cloth when about to be used. If dried and set into one another after being washed, they will become quite clammy. The great objection to using stone milk coolers is the difficulty of drying them thoroughly before being again used. No milk house should be so situated as to admit the steam rising from the boiler which supplies hot water for washing the various utensils; nor should the ground before its windows contain receptacles for filth and dust, but be laid out in grass, or furnished generously with evergreens.

MILK

I have already said that the milk is drawn from the cow into a pail, the size of which may vary to suit the pleasure of the dairymaid. The milk, in being drawn from the cows, is put into a tub and left to cool; but not to become so cold or stand so long as to separate the cream. The tub should be placed in the air and out of reach of animals, such as cats and dogs. After it has cooled, the milk is passed through the milk sieve into the milk dishes, and as much only is put into each dish as not to exceed 2 inches [5cm] in depth.

Iron milk-pan

To know at once the age of milk in the dishes, one mark or score should be made with chalk on the dishes just filled, to show that they contain the last drawn milk, or freshest meal; a second mark is made, at the same time, on the dishes containing the meal before this; and a third is put on the dishes containing the milk drawn before the second meal, and which constitutes the third meal, or oldest milk.

If the cows are milked three times a day, when the first mark is put on the dishes of the evening meal, those of the morning meal of the same day will have 3 marks, to indicate its being the third meal previous, and the dishes of the midday or second meal will have 2 marks. At every meal all the utensils that have been used should be thoroughly cleaned, and set past dry, ready for use when required.

CREAM

The next care of the dairymaid is taking the cream off the milk. In ordinary weather in summer, the cream should not be allowed to remain longer on the milk than three meals; that is, when a fresh meal is brought in, the cream should be taken off the dishes which have three marks, when the milk will be 20 or 22 hours old; but, should the weather be unusually warm, the milk should not be allowed to be more than 18 hours old, or that having two marks, before the cream is taken off it. The reason for using this precaution in taking off cream is that the milk should on no account be allowed to become sour before the cream is taken off, because the cream of sour milk makes bad butter. Let sweet cream become ever so sour after being taken off the milk, and no harm will accrue to the butter. Not that sour cream off sour milk is useless, or really deleterious, for it may be eaten with relish by itself, as a dessert, or with porridge.

Cream skimmer

The cream is skimmed off milk with a thin shallow dish, called a *skimmer* or *creamer*. It may be made of wood or of stoneware and, of the two substances, the ware is preferable for cleanliness, and of ware Wedgwood's or porcelain is the best, being light, thin, hard, highly glazed and smooth. The cream when taken off the milk is put into a *cream jar* (opposite, top) in which it accumulates until churned into butter. Every time a new portion of cream is put into the jar, the cream should be stirred in order to mix the different portions of cream into a uniform mass. The stirring is usually done with a stick kept for the purpose, but spoons of Wedgwood ware are made for doing it. The cream soon becomes sour in the jar, and it should not be kept too long, as it is apt to contract a bitter taste.

Twice a week it should be made into butter, however little the quantity may be at a time. The skimmed milk is put into a tub and made into cheese; but, if a cheese is only made every other day, the milk kept for the following day should be scalded before it is used.

Cream jar

Butter

On converting cream into butter, the first act is to put the churn into a proper state. It is assumed that the churn when last used was put aside in a thoroughly clean and dry state. This being the case, a little hot water, about 2 quarts [2.3 litres], should be poured into it to scald and rinse it.

The dairymaid churns butter as cream settles in Wedgewood dishes behind her

Churning The churning should be done slowly at first, until the cream has been completely broken, that is, rendered a uniform mass, when it becomes thinner, and the churning is felt to be easier. During the breaking of the cream a good deal of gas is evolved, which is usually let off by a small spigot hole if the churn be tight, such as a barrel churn; but in other churns, which have a cover, the air escapes of itself. When the motion of churning is rotatory, it should be continued in the same direction, and not changed backward and forward.

After the cream has been broken, the motion may be a little increased, and

Butter worker

continued so until a change is heard in the sound within the churn, from a smooth to a harsh tone, and until an unequal resistance is felt to be given to the agitators. The butter may soon be expected to form after this, and by increasing the motion a little more, it will form the sooner and, the moment the dairymaid feels that is formed, the motion should cease.

WORKING UP Immediately on being formed, butter should be taken out of the churn and put into the butter tub, one of a broad and shallow form, or a butter worker (shown above) to be worked up. A little cold water being first put into the tub, and the tub set in an inclined position, the butter is spread out, rolled up round the edges, and pressed out by the palm of the hand in order to deprive it of all the buttermilk, for the least portion of that ingredient retained in it would soon render it rancid. The milky water is poured off and fresh poured in, and the butter is again washed and rubbed as often as the water becomes milky. If intended to be kept or disposed of in a fresh state, the large lump is divided and weighed in scales in 1lb [455g] or ½lb [227g] lumps each, and placed separately in the tub amongst water.

Each lump is then clapped firmly by the hand and formed into the usual form in which pounds and half-pounds of butter are disposed of in the part of the country in which your farm is situated. For the table any requisite number of the pounds should be moulded from the lump into prints according to taste, or rolled into forms with small *wooden beaters*, figured or plain (shown left). The made-up butter is then floated in jars with covers, in a clear strong brine of salt and water fit to float an egg, made ready for the purpose.

In churning, the residuum is buttermilk, which, when obtained in large quantity from milk, may be disposed of in towns, or converted into cheese; and, when derived in small quantity from cream, a part may be used for domestic purposes, and the remainder mixed with the food given to the brood sow.

Butter beaters or boards

SALTING If the butter is intended to be salted, it is somewhat differently treated. After being washed clean as above described, it is weighed in the scales, and salt is immediately applied. Practice varies much in the quantity of salt given to butter, so much as from 1oz [30g] of salt to 1lb [450g] of butter, to ½oz [15g] of salt to 1½lb [675g] of butter: 1oz [30g] to 1lb [450g] is too much, it is like curing butter with as little art as a salt herring; ¼oz [7g] of fine pure salt being quite sufficient; and this quantity is intended for keeping butter, for as to powdered butter for immediate use, ½oz [15g] to 2lb [900g] is quite sufficient. In the process of salting, the butter is spread out in the tub and the salt, ground fine, is sprinkled over it by little and little, and the butter rolled up and rubbed down with the side of the hand until the whole mass appears uniform, and is considered to be incorporated with the salt.

To ensure uniform salting, only half the salt should be applied at once, and the butter lumped and set aside until next day, when the other half of the salt should be rubbed in. Whatever of brine or milk may have subsided from the lump in the meantime should be poured off. The salted lump is then put into the jar or firkin on the second day. One great advantage of deferring the making up of butter until the second day is that, without it, the butter will not assume that firm, smooth, waxy texture which is so characteristic a property of good butter.

End-over-end barrel churn

BARREL CHURN In proceeding to the details of the churning machinery, the first class embraces those machines that act by their gyration round a centre, the fluid and the containing vessel revolving together, or partially so; of which the common *barrel churn* may be taken as the type (shown right). The barrel, which is of capacity suited to the dairy, is sometimes provided only with a large square bunghole, secured by a clasped cover, by which it is charged and emptied; while in other cases, one of its ends is movable, and made tight by screwing it down on a packing of

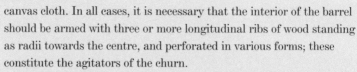

canvas cloth. In all cases, it is necessary that the interior of the barrel should be armed with three or more longitudinal ribs of wood standing as radii towards the centre, and perforated in various forms; these constitute the agitators of the churn.

Each end of the barrel is furnished with an iron gudgeon or journal strongly fixed to it, and to one of them is applied the winch handle by which the machine is turned; while it is supported on a wooden stand, having bearings for the two journals. More than one imperfection attends this construction of churn; from the circumstance of its rotatory motion, it will always have less or more of a tendency to carry the fluid round with the barrel and the agitators, more especially if a rapid velocity of rotation is given to it; and to counteract this tendency, it becomes necessary to reverse the motion at every few turns, which is of itself an inconvenience. There is, besides, the great inconvenience of getting access, either to remove the butter that may adhere to the agitators, or to cleanse the interior of the barrel. This is especially the case where there is no movable end; and even with this convenience for cleaning, the trouble of opening and closing the end is considerable.

Cheese

It is now time to say something on the making of cheese. On a farm of mixed husbandry, as much skimmed milk cannot be procured every day as to make a cheese of ordinary size, but there may be one made every other day.

SCALDING MILK To save skimmed milk from souring in warm weather till the next day, it is necessary to scald it, that is, to put it into a furnace pot and heat it sufficiently, and then let it cool. The fire should be a gentle one, and the milk should be so carefully attended to as neither to burn nor boil, nor be made warmer than the finger can bear. After being thus heated in the morning, the milk should be poured into a cheese tub to await the cheese-making of the following day. The skimmed milk of next morning is poured into the same tub, except about half of it, which is put into the furnace or another pot, and made warm for the purpose of rendering the entire milk of the tub sufficiently

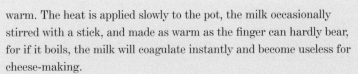

warm. The heat is applied slowly to the pot, the milk occasionally stirred with a stick, and made as warm as the finger can hardly bear, for if it boils, the milk will coagulate instantly and become useless for cheese-making.

On the contents of the tub being mixed by stirring, the rennet or earning is added to the milk, which is allowed to stand some time to coagulate, with a cloth thrown over it, to keep the proper heat.

PREPARING RENNET In the meanwhile, I shall describe the method of preparing the rennet, or reed, or earning. A calf's stomach is usually recommended for this purpose; but as calves' stomachs are not easily obtained in districts where calves are reared, a pig's stomach, which can be easily obtained on every farm, will answer the purpose equally well and, indeed, many believe that it makes the stronger earning of the two.

When the pigs are killed for hams in winter, their stomachs should be preserved for rennet, and they are preserved in this manner. Let the inside skin of the stomachs be taken out; the operation is somewhat troublesome, but may easily be done by an experienced dairymaid. Any curdling in it is thrown away, as being unnecessary, and tending to filthiness, and the skin is then wiped clean with a cloth, not washed. It is then laid flat on a table and rubbed thickly over with salt on both sides, and placed on a dish for four days, by which time it has imbibed sufficient salt to preserve it. It is then hung stretched over a stick near the fire to dry and won [be stored] and, in the dried state, is kept for use as rennet by the next season.

When the rennet is to be used, a strong brine of salt and boiling water, sufficient to float an egg, is made and sieved through a cloth and allowed to cool, to the amount of three imperial pints [1.7 litres] to each skin. One skin is allowed to remain in that quantity of brine in a jar, with its mouth covered with bladder, for three or four days, when the coagulating strength of the brine is tested by pouring a drop or two into a teacupful of lukewarm milk; and when considered sufficiently strong, the brine is freed of the skin, bottled and tightly corked for use. The skin is again salted as before, and spread over a stick to dry and won, and is again ready for use when required. Half a teacupful of this rennet will coagulate as much milk as will make a 15lb [6.8kg] cheese.

Curd mill

CUTTING THE CURD When the milk is sufficiently coagulated, which it will be in half an hour, the curd is cut in the tub with the *curd cutter*, which consists of an oval hoop of copper 9 inches [23cm] long and 6 inches [15cm] wide, and 1 inch [2.5cm] deep, embracing a slip of copper, of the same depth, along its longitudinal axis. The stem of round copper rod rising from each side of the oval hoop unites and after attaining in all 18 inches [46cm] in length, is surmounted by a wooden handle 9 inches [23cm] in length, but 6 inches [15cm] would be enough, by which it is held either by one or both hands, and on the instrument being used in a perpendicular direction, cuts the curd into pieces in the tub. Some people break the curd at first with the hand, but this instrument cuts it more effectively. For those farms that can stretch to the expenditure, a *curd mill* (shown left) can be used to the same effect, although the unbroken curd needs to be fed to the mill by means of a *scoop*.

On being cut, the curd lets out its whey, which is drained off by means of a flat dish being pressed against the *curd cloth* (a linen of open fabric, spread upon the curd). As much of the whey is removed in this way as practicable and the curd will be left comparatively dry, when it receives another cutting with the cutter, and the whey again expressed from it. The curd is then lifted out of the tub and wrapped into the curd cloth, which, in the form of a bundle, is placed upon a drainer lying across the mouth of the tub, and the whey is pressed out of it by main force. The curd becomes very firm after this pressing, and must be cut into small pieces by some instrument before it can be put into the cheese vat. The curd, being made small enough, is salted to taste with salt ground fine.

CHEESE PRESSING After being salted, the curd is put into a *cheesecloth*, spread over a cheese vat and firmly packed into the vat higher than its edge, and, on the curd being covered with the cloth, the vat is placed in the cheese press and subjected to pressure, upon which a quantity of whey will probably exude by the holes in the bottom of the vat. In a short lapse of time, two hours or more, the cheese is turned out of the vat, a clean and dry cheesecloth put in, the cheese replaced into it upside down, and again subjected to increased pressure in the press.

Should whey continue to exude, the cheese must again be taken out of the vat, and a clean cloth substituted; in short, a clean cloth should be substituted, and the pressure increased, as long as any whey is seen to exude; but, if the prior operations have been properly performed, the exudation should cease in about 12 hours, after which the pressure is continued until the press is wanted for a new cheese on the second day.

THE CHEESE VAT The cheese vat, or chesset (shown below), is built in elm staves, as being least liable to burst with pressure, and strongly hooped, and is furnished with a substantial bottom, pierced with holes, to allow the whey expressed to flow away, and a strong wooden cover cross doubled. It is of advantage that the cover fit the vat exactly, and that the vat has as little taper interiorly as possible.

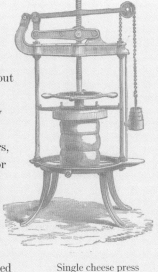

Single cheese press

CHEESE TURNING After the cheeses have been sufficiently pressed, they are put into the cheese room, which should not be exposed to too much heat, drought or damp, as heat makes cheeses sweat; drought dries them too quickly and causes them to crack; and damp prevents them hardening and wonning, and causes them to contract a bitter taste. Cheeses being exposed to a cool, dry and calm air upon the shelves will dry by degrees and obtain a firm skin. The skin becomes harder by being dipped in hot water, but I see no benefit to be derived from such a practice. They should be wiped with a dry cloth, to remove any moisture that may have exuded from them, and turned daily.

Some cheeses burst, and throw out a serous-like fluid, which accident happens in consequence of the whey which was left in it fermenting, and which should have been pressed out. Any cheese that changes the shape that the cheese vat gave it should be suspected of some organic change taking place within it; but if such a cheese does not crack, so as to admit the air into it, it may soon become ripe and mouldy, and prove of fine flavour. The inconvenience of cracks in cheese is the facility afforded to the cheese fly to enter and deposit its eggs; and to prevent their egress, the cracks should be filled up every day with a mixture of butter, salt and pepper, made to a proper consistency with oatmeal.

Wooden chesset or cheese vat

Weeding Corn, Green Crops, Pastures and Hedges, and Casualties to Plants

Summer is the season in which weeds thrive in the greatest luxuriance, and, of course, summer ought to be the season also of the farmer's greatest activity in employing his plans of destruction against them.

Weeding oats

The first crop that requires weeding is the oat, and the weed that most infests it is the creeping plume thistle. The plant should not be cut down before it has attained about 9 inches [23cm], otherwise it will soon spring again from the root, and require another weeding, and by the time it has attained that height, the oats will be about 1 foot [30cm] high. The only implements used in weeding corn are the *hand-draw hoe* and the *weed hook* (right). The hand hoe is used amongst drilled corn, turnips, potatoes and pasture. In using it among drilled corn, the fieldworkers each occupy a row and, on walking abreast in the form of a double echelon, the best worker taking the lead in the centre, the entire ground is cleared of weeds at once, as far as the workers extend across the drills.

The weed hook is used with one hand, the fieldworker walking upright and holding the hook before her near the ground in the position represented in the cut. The stem of the weed is embraced between the

Mud-hoe, harle or claut

clefts of the iron hook by which, it being sharpened to a cutting edge on both sides, the stem is severed by the worker pulling the handle towards her in a slanting direction upwards. The handle is of wood, and about 4 feet [1.2m] in length. The usual way for fieldworkers to arrange themselves when using the weed hook and, indeed, when weeding broadcast corn in any way, is for two workers to occupy every single ridge of ordinary breadth, or four upon the double ridge, each taking charge of the half ridge from the open furrow to the crown. The weeds, when cut over with the weed hook, are left upon the ground and, being young, soon decay.

Weeding barley

Barley is weeded after oats. Besides being troubled with the creeping plume thistle, though not to the same degree as oats, and the charlock, which is pulled when the crop is sown broadcast, and hoed when sown in drills, barley is much infested with the common red poppy, especially in England, and the long smooth-headed poppy.

Weeding wheat

Wheat, though next in order, is not weeded until the stalks have reached the knee, when the principal weed which infests it, the corn cockle or popple, is in bloom. This plant, having a woody stem, is not cut with the weed hook, but pulled by hand.

Weed hook

Weeding hedges

There are a few implements requisite for weeding hedges different from those described for weeding corn and pasture fields. A *hedge spade* is employed, along with a common *Dutch hoe*, for removing the weeds on the hedge bank and a hooked stick draws the weeds, cut over by the Dutch hoe, through between the stems of the thorns into the ditch.

Hedge weed hook

Common spade

The manner of using these implements is this. The hedger steps into the bottom of the hedge ditch, with his face towards the hedge, having in his right hand the crosshead of the hedge spade, and resting its handle in his left hand, works the spade in a horizontal direction, removing all the grassy and other plants growing on the face of the hedge. While the hedger takes the lead of weeding the face of the hedge ditch below the hedge, a fieldworker follows him and removes the weeds with the Dutch hoe along the top and face of the hedge bank behind the hedge.

Weeds growing in the bottom and sides of ditches cause the water to fill up the bottom and break down the sides. The only mode of destroying weeds in ditches is scouring the bottom and paring the edges with the *common spade*, and extracting the roots of the obnoxious plants growing in both.

WEEDING TURNIP AND POTATO

The most useful implement for this purpose is the *scuffler* or *horse hoe*. In the intervals of manual labour in first singling and then hoeing turnips, the scuffler is employed to remove the weeds that may have sprung up between the rows of turnips, and to stir the soil, for the purpose of loosening it for the roots of the plants. The implements required for cleaning the potato crop are the same as for turnips, and the periods and modes of their use are also similar.

Potatoes do, however, require to be set up in all soils because, being tubers occupying the ground below the surface, the earth should be loosened and heaped about them. The *double mould board plough* is an implement essentially requisite in the cultivation of the turnip and potato crop. When duly constructed, it is highly efficient in the formation of the drills or ridgelets for either of these crops. In a variety of forms also, it is much employed in the earthing up of the potato crop.

Turnip drill harrows

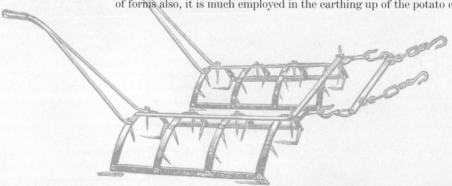

Haymaking

Hay is made both of sown and of natural meadow grasses. The sown grasses are employed for hay in Scotland and, of these, the hay consists of red clover (*Trifolium pratense*) and rye grass (*Lolium perenne*); for, although the white clover (*Trifolium repens*) is sown along with the seeds of the other two, it scarcely forms a part of the first year's grass, and constitutes no part of the hay, which is always taken from the grass of the first year. As hay is thus taken from the first year's grass, it matters not whether the rye grass made into hay is annual or perennial. The annual yields the heavier crop, but the perennial, the finer quality of hay. The natural grasses constitute the hay of England and Ireland. These two sorts of hay are certainly very different in appearance, the sown grasses showing the strong and stiff stems of the red clover and rye grass, and especially when the rye grass is annual, while the hay from the natural grasses is soft and woolly to the feel, and more odorous to the scent, because the sweet-scented vernal grass (*Anthoxanthum odoratum*) always forms a component part.

Haymaking varies according to the means used for conducting it; for if manual labour alone is employed, one process should be adopted, but when mechanical assistance is received, the process should be modified accordingly.

Hay knife

Manual labour

First, then, as to haymaking with manual labour alone. The implements required for the purpose are few and simple. The grass is cut with the *common scythe*. The cutting is either let to labourers by the piece, or the ploughmen of the farm do it, should there be spare time from horse labour between the sowing of the turnips and the hay harvest. The grass will be better and more expeditiously cut down if let by the piece, as the contractors will exert themselves more, and work

Hand hay fork

more hours, than ploughmen, who also have charge of horses, can be expected to do.

On a 500 acre farm, under the five-course rotation, perhaps 20 acres of the 100 acres of new grass will be made into hay. On commencing to cut a field, the direction towards which the clover leans, or should it be thin and upright, the quarter of the wind, which always influences the direction of thin grass, should be attended to; and in both cases the grass should lean away from the mower. It always makes the best work for the grass to be mowed across the ridges. It is fair work for one man to mow one acre every day; and I may here remark, it is no good sign of the weight of the crop if the mowers go over more ground every day. The other implements used in manual haymaking are *forks* (shown left) and *rakes* (shown opposite).

Allowing that the mowers have started early every morning, and there is nothing to prevent them doing so except heavy rain, the grass which had been cut down in the morning should be turned over and shaken up with the forks or, as it is termed, *tedded*, and exposed to the sun and wind, previous to being put into small cocks before the evening.

Cocking

The process for putting it into cocks after the tedding is, for one fieldworker to cast the swathe with a fork from the right open furrow to the crown of the ridge, going in such a direction as to have the ridges upon which the cocks are to be formed on her left hand, and which permits the working of both forks and rakes with the right hand. When the half ridge is thus cleared of the grass with the fork, another fieldworker follows with a rake, and clears the same half ridge of every stem of grass. A third worker follows the rake with a fork and tosses the swathe from the crown to the left furrow of the ridge; and is, in like manner, followed by a fourth worker with a rake, who clears the same half ridge.

On the second or adjoining ridge, a fifth fieldworker throws the accumulating swathe from the furrow to the crown, where her labour is much greater than the workers who wielded the fork on the first ridge, as she has to move the entire grass from the first ridge along with what

Hay hand rake

is found upon the half of the second ridge; and she is followed by a sixth worker with a rake, who clears the ground from the furrow to the crown. In like manner, a seventh and eighth worker put the grass from the crown over the furrow of the second ridge upon the third ridge with the fork, and clear the ground with the rake; and in doing this, the worker who uses the fork is hardest worked of all; but all the rakers have the same degree and extent of labour.

Thus, eight workers are required to clear two ridges of grass, four wielding forks and four rakes, alternately, and the space cleared along the ridges, in this manner, is just the breadth of four swathes of grass, which is, more or less, according to the weight of the crop, but mowers usually cut a breadth of 6 feet [1.8m] at each stroke of the scythe, and each breadth constitutes a swathe. After the second two ridges have been thus cleared, the third ridge being in the middle, contains the grass of five ridges, which is called a *windrow*.

The cocks are raised on this middle or third ridge, and those first made are small, and called grass cocks. They are put together either by the fork or the arms, with narrow bottoms and high in proportion to their breadth, and not exceeding, perhaps, 2 feet [60cm] in height. There will not be room on the ridge, at this time, to put such small cocks in a row, so they may be put up anywhere, as not to crowd upon each other, but afford room for the rakes to clear the ground around them; as it is considered very slovenly work in a hayfield to neglect to clear ground by the rake which had been freed of its grass by the fork. The raking at this time will not occupy above one or two workers, so the rest of the rakers can be employed in assisting the forkers to put together the cocks. The field is left for the night in this state.

Next morning the grass cocks are shaken loosely out on the ridge for exposure to sun and wind; and after this operation is finished, the grass which was cut in the afternoon of yesterday is tedded. In the afternoon the thrown-down grass cocks should be shaken up, after which the grass which had been tedded in the forenoon is windrowed and put into grass cocks, in the manner just described. Before the evening, the thrown-down grass cocks are put into larger cocks, called *hand cocks*, and which are best put together with forks by men. Hand cocks should have small bottoms, built tapered to a fine top about 6 feet [1.8m] in height, and

The 'Albion' mower

placed in a row along the crown of the ridge.

The next morning, the third, the grass cut yesterday afternoon is tedded, and as much more tedded in the afternoon of what was cut early in the morning of the same day, as can be got together into grass cocks before the evening. This is an easy day's work, and reserves strength for the greater labour of the next day, to which all hands of fieldworkers and ploughmen should be collected.

Next morning, the fourth, should it prove a rainy day, let the whole field remain as it was, though the mowers may be able to continue at their work. If fine, toss over first the grass cocks to the sun and air, then ted the small quantity of grass that was mown after the tedding of the previous afternoon and, last of all, throw down and scatter the hand cocks which, by this time, will have subsided considerably. All this will occupy, if not the whole, the greater part of the forenoon, but no more of these respective processes should be undertaken than there is force in the field to put all the hay in cocks before the evening; and of all the processes the tedding of the swathes is the most dispensable. The first thing to be done is to put the hay together that had been scattered, by putting two or three of the hand cocks into one in a row along the crown of the ridge. The hay will be felt to have become much lighter in the hand; for it is surprising how soon hay wins after it has arrived at this stage, if exposed to sun and air.

Let the hand cocks, therefore, be first exposed and then put together at this period, three or four into one, according to the state of the hay and the weather, and the hay will be placed beyond all danger of fermentation and mouldiness. The cocks now assume the name of *ricks* or *colls*, the latter being derived from the French, colline [a small hill]. The colls should be gently tapered to the top, without a projecting shoulder to catch the rain, and its top fastened down with a hay rope, twisted on the spot with the corner of a rake, or with a rope twister or throw crook, taken on purpose to the field, and put across the top of the coll in the direction of the strongest wind to which the locality is subject.

Horse labour

The 'Taunton' haymaker

Let us next consider the making of hay with the aid of horse labour and suitable implements, the employment of which makes a considerable difference in the process. First the hay must be cut and an ingenious machine that has taken the place of the scythe can be found on the most up-to-date farms (shown left). A horse-drawn mower consists of a finger bar within which an oscillating toothed blade passes backwards and forwards driven by the ground wheels. The hay is thus cut most expediently into swathes with little or no need for scythes.

The *tedding machine* is used to ted hay, and which it best does by passing across the swathes, taking up and teasing and scattering them on the ground in the most regular manner.

After the grass has thus been tedded, it is allowed to dry in the sun and wind all the forenoon. In the afternoon the *hay rake*, whether the common *horse rake* (shown right) or *American hay rake*, is employed to rake the tedded grass into a windrow across the four ridges which intervene between every fifth ridge which contains a row of cocks. Where the crop of grass is very thin, the horse rake might carry the grass into a windrow over more than four ridges upon the fifth ridge; but with an ordinary crop it could not perhaps accomplish this, and much less with a heavy crop. After the grass, therefore, has been windrowed across the four ridges, manual labour is employed to put it into grass cocks, as in the case with manual labour. It will be observed that few people, and especially women, are required to conduct haymaking in this way, the heavy part of the duty consisting of making the cocks as often as requisite, which is best done by men.

The 'Star' horse rake

Building a stack

With regard to the stacking of hay, if the entire produce of the field is to be stacked at one time, the colls should be put into a state to stand the weather for a considerable time; two or three being put into one, and the large ricks thus formed are named *tramped pikes*, because they are built and tramped, a man building, and his assistant, a fieldworker, carrying the hay from the fork of the carter and tramping the rick at the same time. Tramped pikes contain from 100 to 150 stones [635–953kg] of hay each, and are commonly placed in a row at the end of the field most convenient for conveying it in carts to the stack in the stackyard, if the hay is retained for the use of the farm or, if disposed of, to a purchaser who stacks it for himself. The reason that hay should be piked if stacked all in one day is that unless hay is in a state to keep, that is, not to ferment in the stack, so much cannot be put together without risk of heating.

The farmers in the south of England employ *rick cloths*, which afford but temporary shelter, but they are quite sufficient to secure the safety of the hay (see illustration below).

A large oblong haystack should be built in this way. In the first place, a dry stance should be chosen, for a damp one will cause the destruction of several stones of hay at the bottom of the stack. The stance should be raised 1 foot [30cm] above the ground, with large stones inscribing the circumference, and the interior filled up with stone shivers or gravel beaten firmly down. Upon this space the stack should be built by two men, who are supplied with armfuls of hay by a number of fieldworkers, whose duty is not merely to carry the hay but to tramp it underfoot in a regular manner from one end of the stack to the other. The two men, each occupying a side of the stack, shake and build up what is called a *dace of the hay* before them as high as their breast, from one end of the stack to the other; and after half its length is built up in this manner, the women go upon it and trample it, and if they hold by one another's hand in a row, their walking will prove the more effective. The breadth of the stack is a little increased to the eaves. The hay is forked from the ground by two or

Mode of erecting a rick cloth over a haystack when being built

three men, and when the stack has attained an inconvenient height for this purpose, there are two or three modes by which hay may be carried to greater height; one is by placing short ladders against the stack, and a man on each, some way above the ground, with his back to the ladder, where he receives the forkfuls of hay from the forker on the ground, and raises the load above his head upon the stack.

A height of 12 feet [3.7m] is enough for the body of the stack, and a breadth of 15 feet [4.6m] is convenient for a haystack, and, with these fixed dimensions, the length may be made more or less, according to the quantity of hay to be stacked. With these dimensions, a new-built stack of 40 feet [12.2m] in length will contain about 2000 imperial stones [12,701kg]. After the body of the stack has attained 12 feet [3.7m] in height, the heading is commenced by gradually taking in the breadth on each side to the ridging, which is elevated half the breadth of the stack above the eaves, and the ends are built perpendicular.

One man and one woman will only find room at the finishing of the top of the stack. A few straw ropes are thrown over the stack to prevent the wind blowing off its new-made top. The stack is left for several days to subside and, unless it has been slowly built and firmly trampled, it may subside in the body to the extent of 2 feet [60cm]. Very probably heat may be indicated in some part of the stack a few days after it is built, by a leaning towards that part, because heating causes consolidation of the hay. A prop of wood placed against the place will prevent the stack subsiding much farther, and the handle of a rake pushed in here and there into the stack will indicate whether the heating is proceeding upwards or to a dangerous extent. A gentle heating will do no harm but, rather, good, by rendering the quality of the hay uniform, and horses do not dislike its effect.

THATCHING

When the hay has fairly subsided, and the heat, if any, is no longer felt, the stack should be thatched; and as a preparatory operation, the sides and ends are neatly pulled straight from angle to angle of the stack, with a small increase of breadth to the eaves. This operation simply consists of pulling out the straggling ends of the hay, which give a rough appearance to the sides and ends, in order to render them smooth; and its use is to save the hay pulled out which would otherwise be bleached useless by exposure to rain, and to prevent rain hanging upon them about the stack.

The heading or thatching consists of straw drawn straight in bundles, held on by means of straw ropes. The thatching should be carried out on both sides of the stack simultaneously by two men, and begun at the same end. The men being mounted on the head of the stack, the bundles of straw are handed up to them on a fork one by one as they are needed, and each bundle is retained in its place on the roof, beside the thatcher, by leaning against a graip stuck into the hay. The straw is first placed over the eaves, handful after handful from the eave to the top of the stack, each length of the straw being overlapped by the one immediately above it. The straw is thus laid from the eaves to the ridge of the stack to a breadth as far as the thatcher can reach at a time with his arms. When the men on both sides meet at the ridge, straw is laid along the stack upon the ridge, to cover the terminal ends, and to support the ropes which keep down the thatch.

When this breadth, of perhaps 3 feet [91cm], or a little more, of the thatch is laid down, its surface is switched down smooth by the thatcher with a supple willow rod, and then a rope is thrown across the stack at its very end, and another parallel to it at 18 inches [46cm] apart, and made fast at both ends, in the meantime, to the sides of the stack. Other ropes, at right angles to the first, are fastened 18 inches [46cm] apart to the hay at the end of the stack and, supposing the side of the roof to be 11 feet [3.4m] along the slope, six ropes running horizontally will be required to cover the depth of the slope, leaving a space of 9 inches [23cm] from the ridge for the place of the uppermost rope, and the rope at each eave is put on afterwards. Each of these horizontal ropes is twisted once round every perpendicular rope it meets, so that the roping when completed has the appearance of a net with square meshes. As every subsequent breadth of thatch is put on, the roping is finished upon it, the advantage of which is that the thatching is finished as it proceeds, and placed beyond danger from wind or rain, or disturbance from after work.

The eave is finished by laying a stout rope horizontally along the line where the roof was begun to be taken in, and twisting it round each perpendicular rope as it occurs; when each perpendicular rope is broken

Haymaking by hand and thrashing ryegrass in the field

off at a proper length and fastened firmly to the hay immediately under the eave, the projecting ends of the thatch over the eave are cut straight along the stack, and give to the heading a pretty finish.

Round stacks

Hay is sometimes built in round stacks, which are kept of a cylindrical form for 7 or 8 feet [2.1 or 2.4m] from the ground, and then terminated in a tapering conical top, and thatched. Such stacks contain from 300 to 500 stones [1,905 to 3,175kg] of hay. This form of stack is convenient enough when a whole one can be brought at once into the hayhouse, but should the stack be of such a size as to be necessary to bisect it perpendicularly, the remaining half is apt to be blown over; or should its upper half be brought into the hayhouse, the under part must be protected by a quantity of straw kept down by some weighty articles and, in such a case, it is seldom that these are put on with sufficient care to keep out rain and resist wind. Upon the whole, the oblong form of stack admits of being most conveniently cut for use, and left at all times in safety; because a section of any breadth can be cut from top to bottom to fill the hayhouse.

SUMMER FALLOWING AND LIMING THE SOIL

Although summer fallow occupies the same division of the farm as green crops – turnips, potatoes and tares – it may most characteristically be regarded as the first preparation for the crop of the following year; it is a transference of a portion of the land, with the labour bestowed upon it, from one year to another; it forms the connecting link between one crop and another. But although the preparation of the soil for a part of the crop of two consecutive years are conducted simultaneously by means of summer fallow, the crops which occupy the soil thus simultaneously prepared are committed to it at very different periods, the green fallow crops being sown early in summer, while the sowing of the fallow crop on the summer fallow is delayed to autumn; so that before the latter makes its appearance above ground, the former have almost advanced to maturity. Since the crop on summer fallow is delayed to autumn – till the eve of commencing another agricultural year – the practical effect of the delay is to dispense with a crop for a whole year on that part of the fallow break which is summer fallowed and, on this account, such a fallowing is commonly called a *bare fallow*. As an entire crop is dispensed with, bare fallowing should impart such advantages to the land as to compensate for the rest and indulgence which it receives.

Bare fallowing, to some extent, is practised every year upon every farm; though the limits of compulsory fallowing have been much circumscribed of late years by the purchase of extraneous manure from distant sources, which are easily conveyed and sold at prices that afford a profit. Those manures, and I only allude to them here, are bone dust and guano. These, added to draining and deep ploughing, have afforded the power to cultivate green crops upon soils which were naturally unfit for them. The land subjected to bare fallowing should have the strongest texture, be foulest of weeds, if any there be, and be situated farthest from the steading, that the carriage of turnips may be rendered as short as practicable. The winter treatment of the fallow land is the same as that for the summer crops, and this has already been described in preparing the soil for potatoes and turnips.

Working the fallow land

When leisure is again afforded to pay attention to the fallow break from the advanced working of the turnip land, the state of the fallow soil should be particularly examined. Should the weeds in the soil consist principally of fibrous and fusiform rooted plants, they will be easily shaken out by the harrows in dry weather; but should the running roots of weeds be found to have threaded themselves through hard round clods, these will not be so easily detached. In such a case, which is of frequent occurrence on strong land, the best plan is to allow the roots to grow for a time, and the force of vegetation will have sufficient power to break the clods, or will render them easily so by the roller, or to reduce the clods by rolling after such a shower of rain as shall have nearly penetrated them.

After such a rolling, the land should be harrowed a double tine, first one way, and then across another way. The weeds and weed roots will then be seen upon the surface. It is not expedient to gather weeds immediately on their being collected by the harrows, as a good deal of fresh soil adheres to them. A day or two of drought should intervene, and the weeds will then be easily shaken free of soil by the hand.

The usual mode of laying dung on fallow land is to feer [mark off] the ridges, employ a tip cart to lay the manure in heaps, spread them and plough the manure in. Fallow land is not dunged so heavily as that for green crops, not so much from fallowed soils not bearing heavy manuring, as from want of manure. From 12 to 15 tons the imperial acre is an ordinary manuring for fallow. The manure need not be so well fermented as for green crops, as there is usually sufficient time for its fermentation in the ground before the wheat is sown.

Liming

While treating of fallow, it is necessary to notice the liming of land, as lime is commonly applied at that period of the rotation of crops, though by no means applied every time the land is fallowed. A week or so before the lime is applied, water is poured on the large heaps of shells, in order to reduce them to a state of impalpable powder. The water will all be absorbed by the lime, which will, nevertheless, continue quite dry, thereby indicating that the water has disappeared by reason of its chemical union with the lime. A great quantity of heat is evolved during the time the lime takes to fall to powder; and when that last has been accomplished, the heaps will have swelled to more than three times their former size, when the lime is said to be slaked, and is then in its most caustic state. While the slaking is proceeding, the land that was manured in drills is cross-harrowed a double tine, to make it flat; after which the ridges are feered [marked off for ploughing]; and the lime is then spread along the feered ridges. The lime is spread in this way. The frying-pan shovels (illustrated on page 159) are the best implements for filling carts with, and spreading lime on land.

Progressively as the lime is spread, ridge after ridge, it is harrowed in and mixed with the soil; and immediately on the entire field being limed, the ridges are ploughed with a light furrow, to bury the lime as little as possible, and which constitutes the seed furrow of the future crop.

It is proper to put a cloth over the horse's back and the harness, and the men may cover their face with crape [a thin, light fabric], to save its orifices being cauterised by the quicklime. The horses, when loosened from work, should be thoroughly wisped down and brushed, to free them of every particle of lime that may have found its way among the hair; and, should the men feel a smarting in their eyes or nose, a little sweet thick cream will be felt as an agreeable emollient; and the same application will prove useful to the horse's eyes and nose.

Building Stone Dykes

It is true that a proper form of stones is a great assistance to the builder of stone dykes, flat thin stones being the best: but flatness and thinness are not the only requisites; they should also have a rough surface by which they may adhere to one another in the wall; and no material, on this account, is so well adapted for the purpose as those derived from sandstone boulders of gravel deposits, which split with the pick into flat stones of requisite thickness when first taken from their matrix and, on being exposed to the air, become dry and hard.

A builder of dry-stone dykes should be brought up to the profession, and when he has acquired dexterity, he will build a substantial wall, at a moderate cost, which will stand upright for many years.

The tools of a dry-stone dyker are few and inexpensive, consisting only of a mason's hammer, a frame as a gauge for the size of the dyke and cords as guides for the straightness and thickness of the dyke. A dyker cannot continue to work in wet or in very cold weather, as handling stones in a state of wetness is injurious to the bare hand; on which accounts, dry-stone dykes are commonly built in summer.

Preparation

The line of fence being determined on, it is marked off with a row of stakes driven firmly into the ground. The upper soil, to the depth it has been ploughed, is removed from the line to form the foundation of the dyke, and it may be driven away, or formed into a compost with lime near the spot for top-dressing grass. After the foundation has been formed by the removal of the earth, the stones should be laid down on both sides as near the line of foundation as practicable, for it is of considerable importance to the builder that the stones be near at hand.

They begin by setting up the frame in the foundation of the proposed line of dyke. The frame is made of the breadth and height of the

Building a drystone dyke

proposed dyke under the cover; and it is set in a perpendicular position by the plummet [plumb bob] attached to it.

A corresponding frame should be placed beyond the point at which the dyke commences, or two stakes, driven into the ground, having the same inclination as the sides of the frame, answer the temporary purpose of an auxiliary frame. Cords are then stretched along the space between the frames, and fastened to each frame respectively, to guide, as lines, the side of the dyke straight, and to gauge its breadth. The frame is held upright and steady by a stiff rail, having the nail projecting through one of its ends, being hooked on to the top bar of the frame, and a stone laid upon its other end.

In building a stretch of dyke, such as the mode above referred to, it is customary to carry up the building at both ends as well as at the middle of the stretch to the levelling of the top, before the intermediate spaces are built up, because those parts being built almost independently, act as pillars in the dyke to support the intermediate building plumb; and they are convenient for pinning the cords into while the intermediate spaces are being built.

LAYING A SCUNCHEON

When the dyke has a scuncheon [corner or angle] for its end, a large boulder should be chosen as the foundation stone; and if no boulders of suitable proportions exist, a large stone should be selected for the purpose, for no better protection can be afforded to the end of a dyke than such a foundation, especially if the scuncheon forms at the same time one side of a gateway to a field. Another boulder, or large stone, should be placed at a little distance from the first, and the smaller stones are used to fill up the space between them, until the space is raised to the height of the boulders.

Laying small stones

There is a great art in laying small stones, and it is, in fact, this part of dyke-building that detects the difference between a good and bad builder. In good dry building, the stones are laid with an inclination downwards, from the middle of the dyke, towards each face. This contrivance causes the rain which may have found its way down through the top of the dyke to be thrown off by both sides; and, to sustain the inclination of the stones, small stones must be packed firmly under their ends in the very heart of the dyke; whereas stones, when laid flat, require no hearting to place them so, and may receive none, to the risk of the dyke bulging out in both faces.

Through-band stones

It tends much to the stability of a dyke to have what is called a through-band stone, a stone that has a face on both sides of the dyke. These should be distributed regularly throughout the wall. The cope, cover or capping stone acts as a through band at the top of the dyke; but in laying the cover, the levelling of the dyke to form its bed should not be made of very small and very thin stones, as is too often the case, as these have little stability, being easily shifted from their position, easily broken and, of course, constantly endanger the safety of both cover and cope.

A stone dyke is in the highest perfection as a fence immediately from the hands of the builder; but every day thereafter the effect of the atmosphere upon the stones, at all seasons, and the accidents to which they are liable by trespasses of individuals, and the violence of stock, render it necessary to uphold their repairs frequently; and this consideration should cause the best-suited materials to be selected for their original erection.

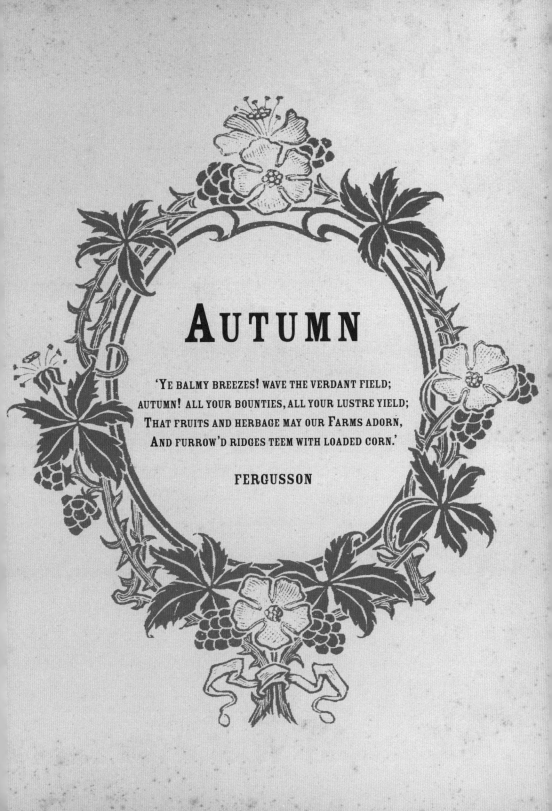

Autumn

'Ye balmy breezes! wave the verdant field;
Autumn! all your bounties, all your lustre yield;
That fruits and herbage may our Farms adorn,
And furrow'd ridges teem with loaded corn.'

FERGUSSON

INTRODUCTION

In contemplating the nature of the different seasons, we have seen Winter, the season of dormancy, in which all nature desires to be in a state of repose, Spring, the season of revival, in which the returning power of nature inspires every created being with new vigour, Summer, the season of progress, in which nature puts forth all her energies, to increase and multiply her various productions, and, now, we see Autumn, the season of completion and of consequent decay, in which nature, in bringing the individual to perfection, makes provision for the future preservation of the kind.

Autumn brings fruition, in which the toilsome labours of the husbandman, for the preceding twelve months, find their reward. The temperature of autumn is high, late August affording the highest average of the year, on account of warmth in the night as well as the day. Such is the heat, that it is no uncommon occurrence for reapers to be seriously affected by it in the harvest field.

The labours of the field partake of the compound character of the season itself. Just as one crop is reaped from the ground, part of the succeeding one is committed to the earth; the autumnal wheat of two successive years being sown and reaped about the same time. The toil endured in harvest is almost incredible. Only conceive the entire bread corn sufficient to support the population of such a kingdom as this to be cut down and carried, in minute portions, in the course of a single month! The usual season of reproduction among the animals of the farm is spring; but the most useful animal of all, the sheep, forms an exception to the rule, autumn being the season in which the ewes are drafted, and the tup is allowed to go with them. There seems in autumn a tendency in the animal frame to disease; sheep are liable to hepatitis; calves to quarter ill [blackleg]; the horse to colic, and even inflammation in the bowels; and stallions and geldings become dull in spirit.

The sports of the field all commence in Autumn. The long-contemplated gatherings in the hills, on the noted 12th of August, in quest of grouse – game, par excellence, of which our country should be proud as its only indigene – cause every shelling to afford shelter to many who, at other seasons, indulge in the far different enjoyments of urban luxuries. Partridge-shooting comes in September, sometimes even before the corn is cut down, and is followed by hare-hunting in October; and after all the fields are cleared of their valuable produce, the inspiring 'music' of the pack is heard to resound through hill and dale.

The great event of Autumn – the harvest – naturally claims a preponderating share of the husbandman's solicitude; and until this important issue of all his toil is secured beyond danger, he cannot rest in quiet. When every straw is safe in the stackyard, and the stackyard gate closed for the season, then, and not till then, is he satisfied of his task being finished, and enjoys undisturbed repose.

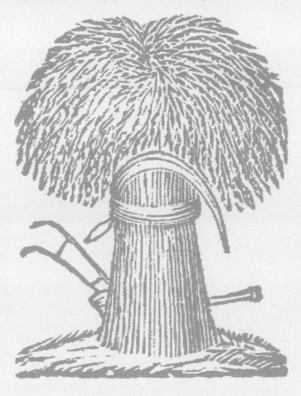

Harvest time

Pulling, Steeping and Drying Flax and Hemp

Flax

WEEDING The only care required by the growing crops of flax in summer and early autumn is weeding, and in its early stage of growth it will be much injured if weeds obtain the mastery. Besides the common weeds which infest the soil, according to its nature, there are others specially found amongst flax; of these, one is the common gold of pleasure (*Camelina sativa*), the seed of which is imported among flax seed, and the plant may be known by its attaining from 2 to 3 feet [60 to 91cm] in height, and having small yellow flowers and very large pouches on long stalks. But a more troublesome weed than this is the flax dodder (*Cuscuta europaea*), inasmuch as it adheres parasitically to the flax plant and, of course, injures its fibre; while the gold of pleasure may be pulled out before the flax is ready.

The weeds, when very young, are picked out by hand from the flax by fieldworkers and, in doing this, the kneeling down upon the flax does it no harm. If weeding be once effectually and timeously [in good time] done, the weeds will not again much trouble the crop.

For later autumn, the pulling, steeping and drying of flax are simple enough, and are processes generally well understood; but Mr Henderson's account of managing the crop, whose sample of Irish flax obtained a gold medal from the Agricultural Improvement Society of Ireland, at their meeting at Belfast in August 1843, being the most practical and, at the same time, succinct I have met with, I shall transcribe it.

RIPENESS And, first, as to test of ripeness, Mr Henderson says, 'I have found the test recommended by Mr Boss to ascertain the degree of ripeness that gives the best produce, with the finest fibre, perfect. It is this: Try the flax every day when approaching ripeness, by cutting the ripest capsule on an average stalk across (horizontally), and when the seeds have changed from the white, milky substance which they first show to a greenish colour, pretty firm, then is the time to pull.'

PULLING FLAX When properly ripened, flax should be pulled in this way. 'I use the Dutch method, say, catching the flax close below the boles; this allows the shortest of the flax to escape. With the next handful the puller draws the short flax, and so keeps the short and the long each by itself, to be steeped in separate ponds. It is most essential to keep the flax even at the root end, and this cannot be done without time and care, but it can be done, and should always be done. The beets should always be small, evenly sized, straight and even, and should never be put up in stooks or windrows, but taken to the pond the day they are pulled.'

STEEPING FLAX Next comes the steeping, which is a most important process, and is the one least understood by growers of flax in this country. If steeping is so long continued as to affect the texture of the fibrous coating, the flax will be injured; and should it not be as long applied as the pithy matter may be easily loosened, much labour will be afterwards incurred in getting quit of it. The water brought to the pond should be pure from all mineral substances, clean and clear.

I put in two layers, each somewhat sloped, with the root end of each downwards. The flax should be placed rather loose than crowded in the pond, and laid carefully straight and regular. I cover with moss sods (from the turf banks) laid perfectly close, the shear of each fitted to the other. Thus covered, it never sinks to the bottom, nor is it affected by air or light. It is generally watered in 11 or 13 days. A good stream should, if possible, always pass over the pond; it carries off impurities, and does not at all impede due fermentation – flood and all impure water should be carefully kept off.

The most particular cause of injury in steeping is exudation of water from the sides or bottoms of the pond. Stripe and discoloration are

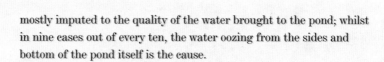

mostly imputed to the quality of the water brought to the pond; whilst in nine cases out of every ten, the water oozing from the sides and bottom of the pond itself is the cause.

DRYING FLAX Great care and neatness are necessary in taking it out. Broken or crumpled flax will never reach the market. Spread the day it is taken out, unless it is heavy rain – light rain does little harm; but, in any case, spread the next day, for it will heat in the pile, and that heating is destructive. Flax 'should be spread even, straight at its length, not too thick, and well shaken, so that there shall be no clots; indeed, if possible, no two stalks should adhere. I rarely let it lie more than five days, sometimes only three. It should never, if possible, be spread on the ground flax grows on, it claps down, and the clay and weeds discolour it; clean lea, or lately cut meadow, is the best.'

Lifting, like all other operations, requires care and neatness to keep it straight to its length, and even at the roots. If the steeping and grassing have been perfect, flax should require no fire; and to make it ready for breaking and scutching, exposure to the sun should be sufficient; but if the weather be damp, the flax tough, and must be wrought off, then it must be fire-dried. The crop of flax, after it is dried, is bulky for its weight; and yields from 3 to 10 cwt. per imperial acre of dried plants.

HEMP

I have also described the sowing of hemp. The crop is pulled and watered, and dried like flax, the weight of produce dressed being little more than flax, from 40 to 45 stones [254 to 286kg] the imperial acre.

The principal use to which hemp is applied is the making of cordage of all kinds, the fibre being both strong and durable. The dried refuse of the stems of hemp, after the fibre has been separated, is used as fuel, and may be converted into charcoal fit for gunpowder. The seed yields about three quarters per acre, and from it is expressed an oil employed with great advantage in the lamp, and in coarse painting. They give a paste made of it to hogs and horses to fatten them; it enters into the composition of black soap, the use of which is very common in the manufacture of stuffs and felts; and it is also used for tanning nets.

Harvesting Rye, Wheat, Barley, Oats, Beans and Pease

We are now arrived at the most important of all field operations – that for which every other that has hitherto been described has been merely preparatory – the grand result, to attain which the farmer feels the greatest anxiety, and which, when attained, yields him the greatest happiness – because it bestows upon him the fruit of all his labour. As harvest work requires a greater number of labourers than usually live on a farm, it is requisite you should hire beforehand a band of reapers on whom you can rely on remaining with you all harvest, and not trust to the chance of a casual supply.

When harvest work goes on in a regular manner through the country, this is an easy and simple mode of conducting harvest work; but should a great proportion of the crop become sooner ripened than was expected, or the weather endanger the safety of the standing crop everywhere, the general demand for hands renders the farmers near towns no better off than those at a distance, for town reapers will then go anywhere for higher wages.

Toothed sickle

Reaping

Every species of grain is cut down with two small instruments, the *scythe* (see page 263) or the *sickle* (illustrated above). The scythe can only be used by men, the sickle by both women and men. Reapers with the scythe must not only be strong men, capable of undergoing great

Everyone was involved in the wheat harvest, women and children included

fatigue, but they must use the instrument dexterously, otherwise they will make rough work and create confusion in the harvest field, where every operation ought to be carried on with precision and least loss of time. The scytheman requires a person to follow him and carefully gather the corn he has mown into sheaves in bands, previously laid down for the purpose, and no person is better fitted for this office than a woman. Another person follows the woman, the *bandster*, whose duty, as his name implies, is to bind the sheaves made by the woman, with the bands he finds lying under them. Another person follows all these, and clears the ground of every loose head of corn with a large rake, and this person may either be a man or a woman; but as one scythesman cannot give sufficient employment to one raker, the economical arrangement is for three scythesmen to work together, with their followers, in what is called a *head*, and one raker will then clear the ground passed over by three scythesmen and their assistants.

There are two ways of reaping with the sickle; in the one, the reapers receive day's wages and, in the other, they reap by the piece. When receiving day's wages, reapers consist of a band of seven persons, called a band icon, who occupy two ordinary ridges, three persons on each ridge and one, called the bandster, to bind the sheaves on each ridge and set them in stooks on one of the two ridges. The bandster should always be a man, as a woman is not able for it; whereas, with a sickle, a woman is as efficient a worker as a man; indeed, what is called a *maiden ridge*, of three young women, will beat a *bull ridge*, of three men, at reaping any sort of corn, on any given day. The six reapers may be all of one sex or the other; but the usual arrangement is one man and two women on each ridge. There is reason for arranging the reapers thus in band wons; because one man can bind the corn cut down by six reapers, and the entire band can cut down and stook two imperial acres of corn every day.

When reapers cut down corn by the piece, each person is paid for what he cuts every day and, to enable the overseer to ascertain that quantity correctly, each reaper, whether man or woman, is put on a ridge by himself. When two or more persons agree among themselves, or a whole family chooses, one ridge is appropriated to them, and the cost of cutting is paid to one of the party.

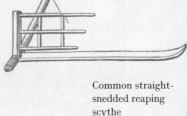

Common straight-snedded reaping scythe

THE SICKLE

Sickles are of two kinds, the *scythe* and the *toothed*. The scythe sickle is so called because of its being provided with a cutting edge, which is kept sharp by the application of sandstone. Its mode of cutting is with a draw of the hand, or with a stroke upon the straw near the root end, in considerable quantity at a time, and the sheaf is gathered with it by rolling the cut corn against the standing with the left hand, occasionally assisting by a push of the sickle.

The toothed sickle, as its name implies, has its edge cut into small teeth which, when applied to straw that is held firm against it, cuts it through like the rasp of a file. From this description, it is obvious that its action is very different from the scythe hook, inasmuch as it cannot cut a straw until the straw is held firm, either directly by the hand, or against a handful of cut corn. Its proper use, therefore, is to cut the corn in small portions. An expert hand can also cut a great quantity of corn with it and, as it requires no sharpening, it occasions no cessation of work.

Whenever the ear is observed to be sufficiently ripe, the crop should be cut down, as the straw will won more rapidly in the stook than standing on the ground. The only matter of doubt, then, in the case is, when the ear is sufficiently ripe. The most ready way of judging of this, in wheat and oats, is the state of the chaff, and of 2 or 3 inches [5 or 7.5cm] of the top of the straw under the ear; if all these are of a uniform straw-yellow colour, and feel somewhat hard in the ear, in the oat, and absolutely prickly to the hand, in the wheat, when grasped, they are ripe; or the grain itself may be examined, and should it feel firm under pressure between the finger and thumb, it is ready for reaping; or should the neck of the straw yield no juice when twisted round by the finger and thumb. Barley should be of uniform yellow colour in the grain and awns, and the rachis somewhat rigid; for as long as it moves freely by a shake of the hand, the grain is not sufficiently ripe, nor will it be of uniform colour. When very ripe, wheat bends down its ear, opening the chaff, and becomes stiff in the neck of the straw, indicating that nature intends the grain to be shaken out.

It may be supposed that, whenever the ear and the entire straw are of uniform yellow colour, the plant is no more than ripe, and so it is; but by the time the straw has fully ripened to the root, the ear will be rigidly bent and ready to cast its seeds with the slightest violence. The same rule may be applied to barley as to wheat, that is, whenever the neck of the straw is ripe, it is time to cut, for when too ripe, the ear bends itself down, diverging the outward row of awns nearly at right angles with the rachis, and is apt to be snapped off altogether by the wind. In regard to oats, the same rule also applies; but there is much less risk of cutting oats unripe, in comparison to allowing them to stand till perfectly ripe, as they are easily shaken out by the wind, the chaff standing apart from the grain. When bean straw turns black, it is fit to cut, and so is pease straw, when the pease become firm in the pod.

It is not easy to describe the best mode of cutting corn with the sickle. In using the scythe hook the body is brought low, by resting chiefly on the right leg doubled under the body, and the left one stretched out to act as a stay to steady the entire frame. The right arm is stretched amongst the corn, and in drawing it towards you, near to, and parallel with, the ground, the standing corn is cut, and is received and held up with the left hand on one side, and by the standing corn behind it on the other. A creeping advance is then made of the body towards the left, and another similar cut made with the sickle, and the additional corn still gathered and supported by the left hand and the standing corn. The great object, in good reaping, is to make short stubble, because more straw is thereby gained to the sheaf, and less left on the field.

In using the toothed hook, the corn is cut in small hookfuls, retained firmly in the left hand, and collected in it as long as it can contain no more, and is then put into the band. In reaping with this instrument, the body bent forward answers the purpose, as small hookfuls can be cut near the ground; indeed, the nearer the ground the easier is the straw cut, but the straw, in this mode of reaping, is always too firmly squeezed in the hand.

THE SCYTHE

Reaping with the scythe is a nice operation and requires considerable skill. The scythes should be mounted and made fit for work some time before being wanted in the harvest field. The various forms of scythes are the *cradle scythe* (right), the common *straight-snedded scythe* (page 261), and that with the bent sned. The greatest favourite amongst mowers is the cradle scythe, because it is easiest to wield by the arms, and does not twist the lumbar region of the body so much as the common scythes.

USING A SCYTHE In commencing to cut a field of corn with the scythe, that side should be chosen from which the corn happens to lie, if it be laid, and if not, then the side from which the wind blows. The scythe makes the lowest and evenest stubble across the ridges, and then also most easily passes over the open furrows. Other things being favourable, it is best to begin at that side of a field which is on the left hand of the mowers. I have already said that reaping with the scythe is best executed by the mowers being in what is called *heads*, namely, a head of three scythesmen, three gatherers, three bandsters and one man raker, or of two scythesmen, two gatherers, two bandsters and one woman raker.

Cradle scythe for reaping

Mowing corn with the scythe

The best opening that can be made of a field for scythe work is to mow along the ridge by the side of the fence, which is kept on the left hand, from the top to the bottom of the field: and while one head is doing this, let another mow along the bottom head ridge, the whole length of the field, and thus open up two of its sides. After this, the first head commences mowing at the lowest corner of the standing corn, across 6 ridges, or 30 yards [27.5m], which is as far as a scythe will cut corn with one sharpening.

SHARPENING THE SCYTHE In sharping a scythe for cutting corn, the scythe stone has to be put frequently in requisition, for unless the edge is kept keen, the mowing will not only be not easy, but bad; and unless a scythesman can keep a keen edge on his scythe, he will never be a good mower, and will always feel the work fatiguing to him.

SCYTHING TECHNIQUE In reaping, it is the duty of the mower to lay the cut corn or swathe at right angles to his own line of motion, and the straws parallel to each other, as at *a a a* (illustrated above); and to maintain this essential requisite in corn reaping, he should not swing his arms too far to the right in entering the sweep of his cut, for he will not be able to turn far enough round towards the left, and will necessarily lay the swathe short of the right angle; nor should he bring his arms too far round to the left, as he will lay the swathe beyond the right angle; and, in either case, the straws will lie in the swathe partly above each other, and with uneven ends, to put which even in the sheaf is a waste of time. He should proceed straight forward, with a steady motion of arms and limbs, bearing the greatest part of the weight of the body on the right leg, which is kept slightly in advance, as seen at *b b b*.

Gathering

The sweep of the scythe will measure about 7 feet [2.1m] in length, and 14 or 15 inches [35 or 38cm] in breadth. The woman gatherers $c\ c\ c$ follow by making a band (see illustration left) from the swathe, and laying as much of the swathe in it as will make a suitable sheaf, such as $d\ d$. The gatherer is required to be an active person, as she will have as much to do as she can overtake. The bandster e follows her and binds the sheaves, and any two of the three bandsters, $f f$, set the stooks g together, so that a stook is easily made up amongst them; and in setting them, while crossing the ridges, they should be placed on the same ridge, to give the people who remove them with the cart the least trouble. Last of all comes the raker h, who clears the ground between the stooks with his large rake of all loose straws, and brings them to a bandster, who binds them together by themselves, and sets them in bundles beside the stooks.

Hand stubble rake

A scythesman will cut fully more than 1 imperial acre of wheat in a day. Of oats, one scythesman will mow fully 2 acres with ease. Nearly 2 acres may be mown of barley.

Binding corn

The bandster, as soon as one band is filled with corn, begins his operations, and he should bind the sheaves in this way. Going to the stubble end of the sheaf, with his face to the corn end, he gathers the spread corn into the middle of the band with both hands and, taking an end of the band in each hand, he moves the sheaf round as much as to

Corn band ready to receive the sheaf

place the corn end parallel to his left arm; then, drawing the ends of the band together as forcibly as he can (especially with the right hand, and as close to the sheaf as possible, keeping the purchase, thus obtained, good with the side of the left hand) he twists the ends of the band round each other with the right hand and doubles the twist under the tightened part of the band, pushing it through as far as to keep a firm hold round the sheaf.

STOOKING

After a sufficient number of sheaves have been bound, the stook is set, and it is set in this way (see illustration below): two sheaves, *a* and its opposite are taken by the bandster, one in each hand, and set a little apart on the ground, with the corn ends close together, such as *b* and *c* are, in such a position that the length of the stook shall stand north and south; and these are intended to form the centre of the stook. Another two sheaves, *d* and its opposite, are set on one side of *a*, in a similar manner, independently, and not leaning against them; another two, *e* and its opposite, are placed in the same manner on the other side of *a*, and so on, till 10 sheaves in a double row of five are thus set.

A sheaf is then split from the band towards the corn end and laid astride upon the top of half the number of standing sheaves, in nearly a horizontal position, and another sheaf *g* is placed in a similar manner upon the tops of the other half number of sheaves, having their butt ends stuck into each other. These last are called hood sheaves, and are intended as a protection to the others against rain, which their drooping position prevents remaining upon.

Barley or oat stook, hooded

Barley and oat stooks are almost always hooded, though, if the weather could be depended on, the precaution would be unnecessary. But as both those kinds of grain remain a considerable time in the field before being fit for the stack, it is the safest plan to put hoods upon the stooks. But wheat stooks (right) are very seldom hooded, because they require to stand only a short time in the field.

Ordinary stook of wheat

Rye

Rye may be reaped or mown in the same manner as the other cereal crops. Its straw, being very tough, may be made into neat, slim bands. It usually ripens a good deal earlier than the other grains; and its straw, being clean and hard, does not require long exposure in the field, and on that account the stooks need not be hooded.

Reaper knife-sharpener

The reaping machine

The almost universal adoption of the reaping machine has rendered it unnecessary for the present race of farmers to acquaint themselves with the working of either the sickle or the scythe, useful as these appliances have been in their day. In all parts of the United Kingdom, and on almost all farms of any considerable size, the reaping machine has superseded the slower and older appliances for cutting down the crops.

Bell's reaping machine

THE EARLY HISTORY OF THE REAPING MACHINE

Although it did not come into extensive use until the middle of the nineteenth century, the reaping machine is by no means a modern invention. It is indeed much older than is generally believed. Both Pliny and Palladius describe a reaping machine worked by oxen, which was much used in the extensive level plains of the Gauls. It is known that before the advent of the nineteenth century several attempts had been made to devise a workable reaping machine. No authentic information has come down to us as to the actual structure of these abortive machines. But soon after the commencement of the nineteenth century, when agricultural improvements were making progress in every direction, and in particular by the extension of the improved machinery to the various branches of farming, active attention was successfully devoted to the invention of the reaping machine. Of the reapers hitherto taken notice of in this work, it is believed that not one of them was ever worked throughout a harvest. Their actual capabilities therefore seem never to have been properly tested. The year 1826 may be held as an era in the history of this machine, by the invention, and the perfecting as well, of the first really effective mechanical reaper. This invention is due to the Rev. Patrick Bell of the parish of Carmylie in Forfarshire (see illustration above). Subsequent makers improved on Bell's machine, and now it exists only as the groundwork of the modern reaper.

MODERN REAPING MACHINES From these small beginnings in the invention and manufacture of reaping machines a great industry has sprung up, from which the agriculture of this country has derived benefits of inestimable value. The reaping machine is now produced in many forms, less or more distinct, suited for different purposes and different conditions of soil and climate. There are the simple mower, adapted merely for mowing hay and leaving it lying as it is cut; the combined mower and reaper, which may be arranged to not only cut the crop, but also to gather it into swathes; the back-delivery, the side-delivery, the self-delivery and the reaper in which the sheaves are turned off by the hand rake. And last and greatest of all comes the *combined reaper and binder*, which is now an established success, performing its intricate and difficult work in a most admirable manner.

USING A HORSE-DRAWN REAPER A day or two before reaping is to be begun, 'roads' should be cut with the scythe all round the field or section about to be cut. Of the two main classes of reaping machines, manual and self-delivery reapers, the former makes better work with heavy or tangled crops. In crops which are moderate in length and not much twisted, the self-delivery reaper, with the saving of one man's labour, is quite as efficient as the manual delivery.

Self-delivery reapers are of two classes, back and side delivery. In the former the sheaves are dropped behind the reaper, while in the latter they are deposited far enough to the side to permit of the machine passing whether the sheaves are tied before it or not. With the side delivery the whole field may be cut without binding any sheaves, while they must be bound as the cutting proceeds if the back delivery or manual machine is used.

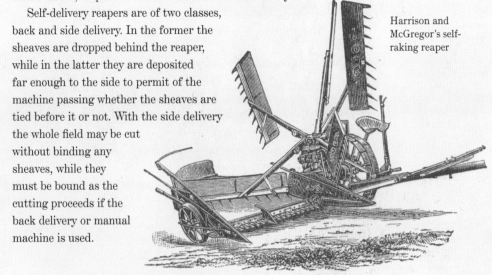

Harrison and McGregor's self-raking reaper

CUTTING Cutting may be done in two ways: either along one side of the field or round about. If the crop is not laid in one direction, the weather is moderately calm, and the crop mostly standing, the roundabout method is the best, as no time is lost returning.

In reaping round about, the binders can be distributed in two ways. If the crop is moderately regular, no better plan need be adopted than that of dividing the circumference of the field into equal divisions for each binder, and sticking a piece of wood into the ground where one division ends and the other begins. By so doing, all unnecessary travelling backward and forward is done away with, and each binder gets a regular share of the good and bad parts of the crop in the field.

By the other system, where each binder has an equal number of sheaves each begins near or where they ended the time previously, the number of sheaves, not the distance, regulating the place. This system is very well suited for comparatively small fields where the number of sheaves allotted to each binder is not very large.

BANDS AND BINDING The corn band is made by taking a handful of corn, dividing it into two parts, laying the corn ends of the straw across each other, and twisting them round so that the ears shall lie above the twist – the twist acting as a knot, making the band firm. The bands should always be made of two lengths of straw, as under no circumstances can a single length of straw be used advantageously. The reaper then lays the band stretched at length upon the ground, to receive the corn with the ears of the band and of the sheaf away from him (see illustration, page 265).

When the band has been laid on the ground, the stubble ends of the straw in the sheaf should be quickly squared by pushing up any straws that are too far down. The sheaf should then be rolled together from the side next to the standing grain, caught firmly in the arms, laid in the band and bound, any lose straws at the cut end being pulled off as the sheaf is thrown to one side.

RAKING A *drag rake* similar to a hay rake, but smaller, is often attached to the machine, which rakes the ground that was cut the swathe previous. In order to allow the rake to work, the sheaves as tied must be conveyed back from the standing grain fully two breadths of the reaper.

Beans

Beans in drills are reaped only with the sickle, by moving backward, taking the stalks under the left arm, and cutting every stalk through with the point of the sickle. When beans are sown by themselves straw ropes are required for bands; but when mixed with pease, the pea straw answers the purpose. Beans are always the latest crop in being reaped, sometimes not for weeks after the others have been reaped and carried.

Pease

Pease are also reaped with the point of the hook, gathered with the left hand, while moving backwards, and laid in bundles, not bound in sheaves, until ready to be carried, and are never stooked at all. In the present times, it may be safely averred that the only means of reaping is by the sickle and the scythe.

CARRYING IN AND STACKING WHEAT, BARLEY, OATS, BEANS AND PEASE

It is necessary that reaped corn remain for some time in stock in the field before it will keep in large quantities in the stack or barn. The length of time will, of course, depend entirely on the state of the weather, for if the air is dry, sharp and windy, the corn will be ready in the shortest time, while in close misty damp, it will require the longest time; but, on an average, for wheat, one week, and for barley and oats, two weeks will suffice.

Mere dryness to the feel does not constitute all the qualities requisite for making new-cut corn keep in the stack. The sap of the plant must not only be evaporated from its outside, but also from its interior; and the outside may feel quite dry, whilst the interior may be far from it; and the knowledge of the latter property constitutes the difficulty of judging whether or not corn will keep in the stack. There is one criterion by which, whether or not a sheaf is fit to keep, may be ascertained with almost certain success, which is, that if the sheaf feel quite dry, the straws be loose and easily yield to the fingers, and the entire sheaf feel light when lifted off the ground, by the hand thrust through the middle of the band, the sheaf is ready; for if it present all these qualities at the band, where it is most compressed, the rest will sure to be won.

While the first-reaped corn is wonning in the field, the stackyard should be put in order to receive the new crop, by removing everything that should not be in it, such as old decayed straw, weeds, implements, etc. Where stathels are used, it should be seen that they are in repair. Straw ropes should also be provided for covering the barley stacks, in case of threatened rain. A *stathel* or *staddle* is a form of stone prop, made up of three parts: a sole, pillar and bonnet. On these sits a framing of wood which forms the base of the stack.

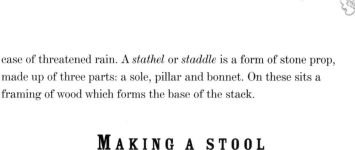

MAKING A STOOL

A stool of straw for a stack is made in this way. Stick a fork upright in the ground in the centre of the intended site of the stack. Put a quantity of dry straw around the fork, and shake it up as you would do the litter of a horse, spreading it out to about the size and form the stack should occupy on the ground. Take a long fork, with the radius of the stack marked clearly upon its shaft, embrace the upright fork firmly between its prongs and, holding its shaft at the specified distance of the radius, push in or pull out the straw with your feet into the shape of the circle described by yourself in walking round the circumference of the stool.

LOADING THE CART

A corn cart is loaded with sheaves in this way. The body is first filled with the sheaves lying with their butt ends towards the shaft horse's rump at one end, and the back end of the cart at the other. When the corn is on a level with the frame or shilments of the cart, the sheaves are then laid across the body of the cart in a row along both sides of the frame, with the butt ends projecting as far beyond the frame as the band, the sheaf on each corner of the frame being held in its place by being transfixed upon a spike attached to it. Another row of sheaves is placed above the first, and the corner ones kept in their places by a wisp of corn, held fast by the band being placed under the adjoining sheaf. Sheaves are then placed along the cart with their butt ends to both its ends, in order to hold in the first-laid sheaves, and to fill up the hollow in the middle of the load. Thus row after row is placed, and the middle of the load filled up till as much is built on as the horses can conveniently draw, 12 large stooks being a good load. A load thus built will have the butt ends of all the sheaves on the outside, and the corn ends in the inside.

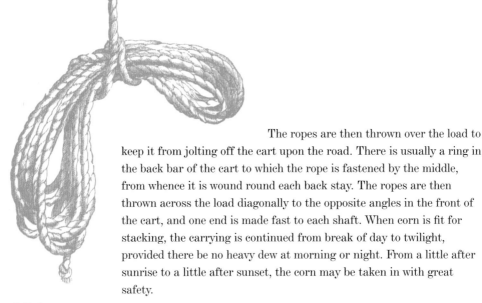

Coiled-up cart rope

The ropes are then thrown over the load to keep it from jolting off the cart upon the road. There is usually a ring in the back bar of the cart to which the rope is fastened by the middle, from whence it is wound round each back stay. The ropes are then thrown across the load diagonally to the opposite angles in the front of the cart, and one end is made fast to each shaft. When corn is fit for stacking, the carrying is continued from break of day to twilight, provided there be no heavy dew at morning or night. From a little after sunrise to a little after sunset, the corn may be taken in with great safety.

BUILDING THE STACK

In setting a loaded cart to the stool or stathel of a stack, it should be studied to let the ploughman have the advantage of any wind going in forking the sheaves from the cart. The stack should be built in this way. Set up a couple of sheaves against each other in the centre of the stathel, and another couple against them in the other direction. Pile others against these in rows round the centre, with a slope downwards towards the circumference of the stathel, each row being placed half the length of the sheaf beyond the inner one, till the circumference is completed, when it should be examined, and where any sheaf presses too hard upon another, it should be relieved, and where there is slackness, another sheaf should be introduced. Keeping the circumference of the stack on the left hand, the stacker lays the sheaves upon the outside row round the stack, putting each sheaf with his hands as close to the last set one as he can get it, and pressing each sheaf with both his knees. When the outside row is thus laid, an inside one is made, with the butt end of the sheaves resting on the bands of the outside row, thereby securing the outside sheaves in their places and, at the same time, filling up the body of the stack firmly with sheaves. A few more sheaves may be required as an inmost row still, to fill up and make the heart of the stack its highest part.

The number of rows required to fill the body of a stack depends on the length of the straw and the diameter of the stack. In crops of ordinary length of straw, such as from 4½ to 5 feet [1.4–1.5m], a stack of 15 feet [4.5m] diameter is well adapted; and one inside row, along the bands of the outside one, with a few sheaves laid across one another in the centre, will make the stack completely hearted. As every cart is unloaded, the stacker descends to the ground, by means of a ladder, and trims the stack, by pushing in with a fork the end of any sheaf that projects further than the rest, and by pulling out any that may be too far in.

Ladder

The ladder is an indispensable implement for the builder of stacks; not only to allow him to get upon and from the stack when he chooses, with ease to himself, and safety to the stack, but to enable him to measure the height of its eaves, so that all the stacks of the same diameter may be built of a uniform height. A couple of 10 foot [7.3m], a couple of 15 feet [4.6m], and a 24 feet [27.5m] ladder will suffice for all the purposes of the stackyard, and repairs of the roofs of the steading and houses. As the stack rises in height with cartload after cartload, the trimming cannot be conveniently done with a fork; a thin flat board about 20 inches [51cm] in length, and 10 inches [26cm] broad, nailed firmly to a long shaft, is an appropriate instrument for beating in the projecting ends of the sheaves, and giving the body of the stack a uniform roundness (illustrated top right).

Building a stack of corn

Stack trimmer

MAKING THE EAVES

A stack of 15 feet [4.6m] in diameter should ultimately stand 12 feet [3.7m] high in the leg, and an allowance of 1 foot [30cm], or 1½ feet [46cm], for subsidence, after the top is finished, according to the firmness of its building, is generally made. The height is measured with the ladder, and allowing 2 feet [60cm] for the height of the stathel, a 15-foot [4.6m] ladder will just give the required measure of the height of the leg before the top is set on. The eaves of the stack are formed according to the mode in which it is to be thatched. If the ropes are to be placed lozenge-shaped, the row of sheaves which forms the eaves is placed a little within the topmost outside row, and after the top is fully finished, its slope will be the same as that of a roof, namely, 1 foot [30cm] less of perpendicular height than half the diameter. In finishing the top, every successive row of sheaves is taken as much farther in as to give this requisite slope, and the bevelled bottoms of sheaves, as they stand in the stock, answer this purpose well; the hearting being particularly attended to in every row, till the space in the centre of the stack is limited to an area upon which four sheaves can stand with their tops uniting, and their butt ends spreading out to give a conical form to the top; and these sheaves are kept firm in their place against gusts of wind with a straw rope wound round them and fastened to the sheaves below.

Before thatching can be carried on, preparations should be made for it some time before, that is, straw should be drawn in bundles, and ropes twisted ready to be used; and a rainy day in harvest cannot be better appropriated than to such purposes.

Throw crook

STRAW-ROPE MAKING

Straw is twisted into rope with different instruments, and in different styles. The simplest instrument is the old-fashioned *throw crook* (above left) or the *improved throw crook* or *wimble* (left). It is used in this way. The left hand holds by the ring at the end of the shank; and round the point of the head is received a wisp of straw from the person who is to let it out to be spun. The right hand holds the middle of the shank

Improved throw crook or wimble

loosely, and causes the head to revolve round an axis formed by the imaginary line between the head and ring, and the twister walks backward while operating with the instrument.

The person who lets out the straw sits still on a stool, or on bundles of straw, and, using the left hand nearly closed, restrains the straw in it till sufficiently twisted, and then lets it out gradually, while the right hand supplies the straw in equal and sufficient quantities to make the rope equal throughout, the twister taking away the rope to the requisite length as fast as the spinner lets it out (shown below). The spinner then winds the rope firmly on his left hand in an ovoidal ball, the twister advancing towards him, as fast as the spinner coils the rope, with a hold of the end which secures the ball firm.

The best sort of straw for rope is that of the common or Angus oat, being soft and pliable, and it makes a firm, smooth, small rope. An ordinary length of a straw rope may be taken at 30 feet [9m]. Straw ropes can be twisted in quite a different way, with a machine similar to the one used by ropemakers to twist their cords. In using it the twister sits still, while the spinners carry the straw under their arm, and move backwards as they let out the straw. The spinners then wind the rope upon the left hand, and advance, during the winding, towards the machine, where they are ready to begin to spin again. Usually three spinners let out to one twister, and as they can spin as fast with this machine as with the crook, the cost of making each rope will be less than that given above; but an inconvenience attends the use of this twister – when one of the spinners breaks his rope, he is thrown out of work till the others begin a new rope; and all the spinners must let out with the same velocity, otherwise one will make a longer or a harder twisted rope than the others.

Making a straw rope with a throw crook

Thatching

Having the materials ready – drawn straw and straw ropes – the covering or thatching of a stack is done in this manner. On the thatcher ascending to the top of the stack by means of a ladder, which is immediately taken away, a bundle or two of straw is forked up to him by his assistant, a fieldworker, and which he keeps beside him behind a graip, as noticed in covering the haystack. The straw is first laid upon the eaves, beyond which it projects a few inches, and then in an overlapping manner upwards to the top. Where a butt end of a sheaf projects, it should be beaten in, and where a hollow occurs it should be filled up with a little additional straw. In this manner the straw is evenly laid all round the top of the stack to the spot where the thatcher began.

Suppose he has laid the covering on the top of the stack all round to the line from a to b (see illustration), before closing up with which he makes the top a, consisting of a small bundle of well-drawn long straw, tied firmly at one end with a piece of cord; the tied end is cut square with a knife, as shown at a, and the loose end is spread upon the covering, and forms the finishing to it. To secure the top in its place, a straw rope is thrown down by the thatcher from a to d, the end of which his assistant on the ground fastens to the side of the stack. After passing the other end of the same rope round the top, he throws it down in the same direction, where it is also fastened to the stack. In like manner, he throws down both the ends of a rope from a to c, where they are also fastened by the assistant. These two ropes are seen at e and f. Having thus secured the top, the thatcher closes in the covering from a to b, when the ladder is placed to receive him.

Taking the ladder to c, he inclines its top over the covering of the stack, and secures its lower end

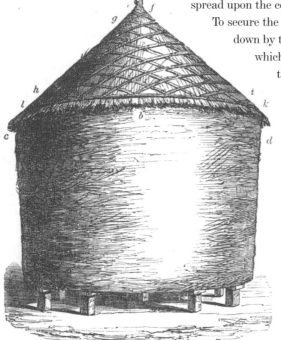

Lozenge mode of roping the covering of a corn stack

from slipping outwards by a graip thrust against it into the ground. He then mounts and stands upon the ladder at the requisite height above the eaves at c, and there receives a number of ropes from his assistant, which he keeps beside him, between the ladder and the stack. Holding on by the end of a coil of rope, he throws the coil from where he stands at c down in the direction of d, to his assistant, who, taking it in hand, allows the thatcher to coil it up again upon his hand, without ruffling the covering of the stack, till of sufficient length to be fastened to the side of the stack. The thatcher then throws the other newly coiled end in the same direction of d where, on his assistant taking hold of it, the thatcher retains the rope in his hands by the double, and places it in its position at g, a little way below e, and keeps it in its place till the assistant pulls it tightly down, and makes it fast to the stack like the other end. Thus the thatcher puts on every rope below g, till the last one on that side has reached h. He then takes the ladder to d, where he puts on every rope below f till they reach the last one, i. Ropes thus placed from opposite sides of a stack cross each other in the diamond or lozenge shape represented in the illustration. It will be seen that a windy day will not answer for laying on the covering of stacks.

To give the thatch straw a smoothness, it should be stroked down with a long, supple rod of willow; and to give the ropes a firm hold, they should receive a tap here and there with the fork, while the assistant is pulling the last end tight. The thatcher is obliged to throw down the rope at first coiled, and to coil up again the second end before it is thrown down, because the loose ends of straw ropes would not descend within reach of the assistant. The ends of the ropes are fastened to the stack by pulling a handful of straw from a sheaf a little out of the stack, and winding the rope round it, and the knot, thus formed, is pushed between the rope and stack and keeps the rope tight.

I shall describe the finishing process at once. An eaves rope, $k\,l$ (see illustration), is spun long and strong enough to go round the stack. Wherever two ropes from opposite directions cross at the eaves rope, they are passed round it, and being cut short with a knife, are fastened to the stack, immediately below the projecting part of the thatch over the eaves. Thus the two ends of all the 20 ropes are fastened to the stack, and the thatch is cut with a knife round the eaves, in the form shown from d by b to c.

DRAFTING EWES AND GIMMERS, TUPPING EWES, AND BATHING AND DIPPING SHEEP

When last speaking of sheep, the lambs were weaned and buisted [marked]. One of the processes amongst sheep in early autumn, in the beginning of August, is drafting ewes and gimmers [young females], that is, separating those to be disposed of from those to be kept. Drafting, however, applies only to a standing flock of ewes. By a standing flock is meant a fixed number of ewes, which are made to rear a lamb every year.

There are various marks of deterioration which determine the drafting of ewes: bareness of hair on the crown of the head, which makes them obnoxious to the attacks of fly in summer; deficiency in eyesight, which prevents them keeping with the flock, and choosing out the best parts of pasture, and best points of shelter; ill-shaped teeth and jaws, which disable them from masticating their food so well as they should; want of teeth from old age, when, of course, they cannot crop sufficient food to support their lambs; hollow neck, which indicates breeding too near akin; hollow back, which implies weakness in the vertebral column, thereby rendering them unfit to bear lambs to advantage; flat ribs, which confine the space for the foetus within the abdominal region; a drooping, tall head, which affects the length of the

A bathing can or flask

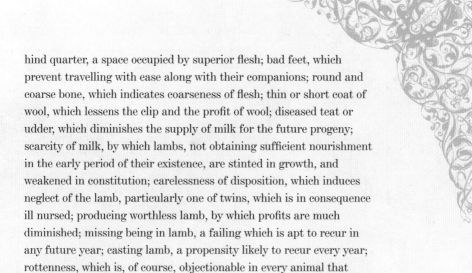

hind quarter, a space occupied by superior flesh; bad feet, which prevent travelling with ease along with their companions; round and coarse bone, which indicates coarseness of flesh; thin or short coat of wool, which lessens the clip and the profit of wool; diseased teat or udder, which diminishes the supply of milk for the future progeny; scarcity of milk, by which lambs, not obtaining sufficient nourishment in the early period of their existence, are stinted in growth, and weakened in constitution; carelessness of disposition, which induces neglect of the lamb, particularly one of twins, which is in consequence ill nursed; producing worthless lamb, by which profits are much diminished; missing being in lamb, a failing which is apt to recur in any future year; casting lamb, a propensity likely to recur every year; rottenness, which is, of course, objectionable in every animal that produces young; shortness of breath, which prevents them seeking their food, and eating so much of it, as they should; tendency to scouring [diarrhoea], or the opposite, the former imposing weakness, the latter inducing inflammation; delicacy of constitution, which disables them from withstanding the ordinary changes of the weather; diminutive stature, or inordinate size, which destroys the uniformity of the flock. This is a long list of faults incidental to ewes, and yet every one may be observed, and which every breeder of sheep is desirous to get rid of.

It is not in the power of the breeder to draft every ewe having an objectionable property every year, because, the farm supporting a stated number of ewes, the extent of their draft depends on the number of good substitutes which may be obtained from the gimmers; for it is obvious no good object is attained by drafting a bad ewe, and taking in its stead a bad gimmer. The number of gimmers fit to be transferred to the ewe flock should therefore, in the first instance, be ascertained, and a corresponding number of the worst ewes drafted. In drafting gimmers, many of the above faults may be observed in them also. Thus ample drafting can alone ensure a sound, healthy, well-formed, young and strong-constitutioned flock of ewes.

The draft ewes and gimmers should be put on the best grass the farm affords, immediately after being drafted and buisted, to fatten as soon as possible, as they are usually sold before the time arrives for putting lean sheep on turnips. Drafts should be got ready for sale in September.

Blackfaced ram

Tups

Tups require but little preparation on being put amongst ewes. If their skin is red in the flanks when the sheep are turned up, they are ready for the ewes, for the natural desire is then upon them. Their breast, between the fore legs, is rubbed with keil or ruddle, on being placed amongst ewes, that they may mark their rump on serving them. It is the duty of the shepherd to mark every ewe so served, that he may observe whether or not the season returns upon her, and to be prepared for the day of her lambing when it arrives. The period of gestation of the ewe is five months and, as the tup is usually put amongst the ewes from the 8th to the 11th October, the first lambs may be expected to appear on the 8th or 11th of March following. A young, active tup, a shearling, will serve 60 ewes, an old one, 40 ewes.

Tups are not selected for ewes by mere chance, but according as their qualities may improve those of the ewes. When ewes are nearly perfect, they may be selected for breeding tups. A good ewe flock should exhibit these characteristics: strong bone, supporting a roomy frame, which affords space for a large development of flesh; abundance of wool of good quality, keeping the ewes warm in inclement weather, and ensuring profit to the breeder; a disposition to fatten early, which enables the breeder to get quit of his draft sheep readily; and being prolific, which increases the flock rapidly, and is also a source of profit.

Now, in selecting tups, it should be observed whether or not they possess one or more of those qualities, in which the ewes may be deficient, in which case their union with the ewes will produce in the progeny a higher degree of perfection than is to be found in the ewes themselves, and such a result will improve the state of the future ewe flock.

After three weeks have elapsed from putting the tup amongst the ewes, he should be withdrawn; because lambs begotten so long after the rest will never coincide with the flock. Tups should, after serving, be put on good pasture, as they will have lost a good deal of condition, being indisposed to settle during the tupping season. The ewes [mature female sheep] and gimmers [young female sheep] may now be classed together, and get such ordinary pasture as the farm affords.

Bathing

Immediately after the arrangements for tupping the ewes are made, part of the sheep stock undergo a preparation for being put on turnip, and the preparation consists of bathing them with a certain sort of liquid.

Sheep are affected by a troublesome insect, the *keb* or *ked*, or *sheep tick*, which increase so much in numbers, as the wool grows towards autumn, as to become troublesome to sheep; and were means not taken to remove them, the annoyance they occasion the sheep would cause them to rub themselves upon every object they can find, and in thus breaking their fleece, deteriorate its value to a considerable extent. Another reason for bathing sheep is that, on experiencing so great a change of food as from grass to turnips, cutaneous eruptions are apt to appear on the skin, even to the extent of the scab, which would deteriorate the fleece even more than the rubbing occasioned by the ked. The materials used in the bath are *tobacco*, *spirit of tar*, *soft soap* and *sulphur vivum*, and a useful implement in bathing sheep is the *bathing stool* (shown right).

Bath stool for sheep

The bathing is conducted in this way. A sheep is caught and placed on the stool on its belly, with its four legs through the spars, and its head towards the seat *a*, on which the shepherd sits astride. The wool is shed [parted] by the shepherd, with the thumbs of both hands, from one end of the sheep to the other, and when he has reached the farthest end of the shed, an assistant, a fieldworker, pours the liquor from the flask equally along the shed, while kept open by the shepherd with both hands. The sheds made are one along each side of the backbone, one along the ribs on each side, one along each side of the belly, one along the nape of the neck, one along each side of the neck, and one along the counter. From these sheds the bath will cover the whole body.

Bathing sheep

The shepherd and his assistant will bathe 40 sheep in a day in this manner. Dry weather should be chosen for bathing, else the rain will wash away the newly applied bath; and coarse clothes should be worn by those who administer the bath, as it is a very dirty process.

Dipping

Instead of bathing sheep in this manner, which is the old-fashioned one, it has of late years been recommended to dip them in tubs containing liquids, which answer the same purpose as the bathing materials I have spoken of. It is obvious that a liquid, to be applied with certainty to the entire body of the sheep through its wool, must be as limpid as water; and, accordingly, all the dipping compositions are dissolved in large quantities of water.

The dipping apparatus consists of a box for holding the bath, to one side of which is attached a drainer, upon which the bathed sheep are placed, and the liquid squeezed out of their wool. From the drainer the sheep are slid down a short inclined plane into a pen in which the bath drips from them, and runs away by a channel, and is collected in a vessel for use again.

The sheep are dipped in this way. Every sheep is held by two men. One holds the head with the right hand, and the two forefeet with the left; the other man holds the two hind legs. They dip the entire body of the sheep, with the back undermost, in the bath, with the exception of the head. They then place the sheep upon the drainer, squeeze some of the bath out of the wool with the hand, and this part of the bath returns immediately into the box. The sheep is then slid down on its side on the inclined plane into the pen, where it remains for a time to drip its wool of superfluous bath.

Lifting and Pitting Potatoes

The harvest work of a farm cannot be said to be completed until the potato crop is taken out of the ground and secured against the winter's frost. By October the potatoes may be expected to be ready for lifting. The fitness of potatoes for lifting is indicated by the decay of the haulms [stems]; for as long as these appear at all green, you may conclude the tubers have not yet arrived at maturity. Immediately after the fields are cleared of corn, the potatoes should be taken up and secured, to allow the land to be ploughed up for wheat.

There are two modes of lifting potatoes, namely, with the *plough* and with the *potato graip*. With either instrument a large number of people to gather the potatoes are required, each of whom should be provided with a small semi-spheroidal shaped basket with a bow handle (see page 165), to gather the potatoes into, and then to put them into sacks or close-bodied carts. When a farmer lifts potatoes on his own account, they are usually put into a cart and carried direct to the pits. When he lifts them on account of a purchaser, or a number of purchasers, they are measured on the spot from the basket and put into sacks, in which they are easily delivered.

The potato plant – tubers reach maturity when the flowers and plant have completely died back

Potato-separator

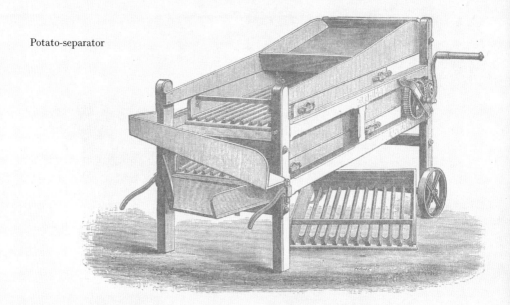

USING THE PLOUGH

In employing the plough to take up potatoes, the common one, with two horses, answers well; but as the potatoes run the hazard of being split by the coulter when it comes in contact with them, it should be taken out, the sock being sufficient to enter the plough below the drill, and the mould board to turn them out of it. The plough in going up splits one drill, and in returning splits the next, but no faster than a band of gatherers, of fieldworkers, if numerous enough, but if not, assisted by hired labourers, can clear the ground of them into the baskets.

Potato gathering should not be continued so late in the evening as the tubers cannot be easily seen; nor should it be prosecuted in wet weather, which causes the

Potato raiser attached to a plough

earth to adhere to them, and renders them undistinguishable from the earth itself. Of course every one, the smallest tuber, should be taken off the ground, not only on the score of economy, to realise the whole crop, but to remove them as a weed from among the succeeding crop. After the field has been gone over in this manner, the harrows are passed a double tine to bring concealed tubers to the surface, when they are gathered by the people, and to shake the haulms [stems] free of soil. These after gathered potatoes are usually reserved for pigs and poultry. Whenever the field is cleared of the crop, the haulms are gathered by the fieldworkers and carried to the compost stance, to be converted into manure, and these are the only return which the potato crop makes to the soil.

A simple instrument, which may be substituted in the plough for the mould board, for turning potatoes out of the drill, is the *potato raiser* or *brander* (shown below left). This is attached to the right side of the head and stilt of a plough, in lieu of the mould board by screws, being placed close behind the sock. The mode of operation of the brander is, that while the earth partly passes through it, and is partly placed aside by it, the potatoes are wholly laid aside, so there are few of them but are left exposed on the surface of the ground.

USING THE GRAIP

When potatoes are taken up by manual work, it is done with the *potato graip* (shown right), the prongs of which are flattened. Being rather severe work to use this graip, men are employed for the purpose, one man taking one drill, close beside that of his fellow workmen, while two gatherers to every man are ready to pick up the potatoes he turns out into the baskets. In using the graip, it is inserted into the side of the drill, and below the potatoes, with a push of the foot, and the graipful of earth thus obtained is turned on its back into the hollow of the drill, exposing the potatoes to view on the top of the inverted earth, from whence they are gathered. The men then pass the prongs of the graip here and there through the inverted graipful and the soil on the drill, to detect and expose to view every tuber lurking beneath the soil.

Potato graip or fork

Storing

In regard to the storing of potatoes, there is no difficulty in the early part of winter, when a low temperature prevails, and vegetation is lulled into a state of repose. Potatoes may therefore be kept in almost any situation in the early part of winter; but then, if damp is allowed to surround them for a time, it will inevitably rot them, and if air finds easy access to them at all times, the germ of vegetation will be awakened in them at the first call of spring. To place potatoes beyond the influence of those elements as long as convenience suits, they should be stored in a dry situation, and be covered up from the air; and no mode of storing affords most ready means for both those requisites than the ordinary forms of pits in dry soil.

There are two different forms of ordinary potato pits, the one being conical, the other prismatic in shape (see illustration). The conical form is usually employed for pitting small quantities of potatoes, and is well suited for small farmers and cottars; the prismatic is the form commonly adapted for storing large quantities. For both sorts, a situation sheltered from the north wind should be selected, and the ground should either be so dry of itself as to absorb the rain as it falls, or so inclined as to allow surface water to pass away quickly from the site of the pits. The site should be conveniently situated for opening the pits and admitting carts to them, and so near the corner or side of a field as not to interfere with its being wrought in winter.

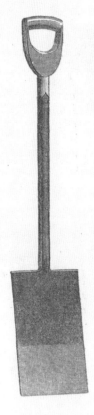
Spade

A conical pit of potatoes is formed in this manner. If the soil is of ordinary texture, and not very dry, let a small spot of its surface be smoothened with the spade. Upon this spot let the potatoes, as they are taken out of the cart, be built by hand in a conical heap, not exceeding 2 feet [60cm] in height; and the breadth which a cone of that height will occupy, so as not to impose much trouble in piling up the potatoes, will not be less than 4 feet [1.2m], and is more likely to be 5 feet [1.5m]. The potatoes are then covered with a thick thatching of dry, clean straw. Earth is then dug with a spade from the ground in the form of a trench around the pit, the inner edge a, of the trench being as far from the pile of potatoes as the thickness of the covering of earth to be put upon it, which is considered sufficient at 1 foot [30cm]. The first spadeful is laid around the potatoes on the ground, and the earth chopped fine and beaten down with the spade, in order to render the earth as impervious to cold as possible, and the drier the earth is, the less effect will frost have upon it, and the less distance will it penetrate through it. Thus spadeful after spadeful of the earth is taken from the trench and heaped on the straw above the potatoes, until the entire cone $a\,b\,c$ is formed, which is then beaten smooth and round with the back of the spade. The top of the cone at b will then be about 3 feet 3 inches [99cm] in height, and the width of the cone from c to a about 7 feet [2.1m]. The trench round by $a\,c$ should be cleared of earth that no surface water may lay near the pit, and an open cut should be formed from the lowest side of the trench to allow the water to go away most freely.

The prismatic or long pit $d\,e\,f\,g$ (also illustrated) is formed exactly in the same manner, with the exception that the potatoes in it are piled in a straight line along its sides $d\,h$, instead of round, as in the case of the conical pit $a\,b\,c$. The height of the pile of potatoes should not exceed 2½ feet [75cm], and its breadth will spread out to about 7 feet [2.1m], and allowing 15 inches [38cm] for thickness of straw and earth, the height of the finished pit will be 3 feet 9 inches [114cm], and breadth 9 feet 6 inches [2.9m]. The direction of a long pit should always be north and south, in order to place both its sides within reach of the sun's rays.

Conical and prismatic potato pits

Sowing Autumn Wheat

How ceaseless, indeed, is the round of rural labour! No sooner does the farmer secure his crop, the progress of which towards maturity has excited his most lively solicitude during the course of a whole year, than he begins to sow the succeeding one, and strives to prepare as much land for it before winter sets in as he possibly can secure. The crop usually sown in autumn is wheat.

You will recollect that we left off the working of summer fallow after the land was dunged, and when the land was to receive no lime. It is now our business to finish the summer fallow, by the sowing of autumn wheat, the crop for which the land was specially prepared by fallow. The first process is the levelling of the drills which cover the dung, by harrowing them across a double tine; and, unless the land is of very strong clay, one double tine will be sufficient for the purpose. After the land has been harrowed level, any root weeds that have been brought to the surface should be removed, but surface weeds will soon wither in the sun and air. The land should now be *feered*, to be gathered up into ridges.

When land is naturally strong enough to grow wheat, but is somewhat soft, and so wet below as to make it apt to throw out the wheat plant in spring, the best plan is to make a seed bed by ribbing with the *small plough*. The wheat can be sown broadcast over the ribs, and harrowed in with a double tine along. The ribbing gives the wheat a deeper bed in the soil than mere harrowing, and a deeper

Corn and seed drill

hold of the soil in spring, and it has also the advantage of stirring only the dry surface soil for the seed bed. It can only be practised, however, on land that has been ridged up for a seed bed for a considerable time, as the small plough does not make good work on new-ploughed land, it, small as it is, going too deep, and making the drills too wide; and it is never employed on land that has not been ridged, being unfit to turn up land in a hardened state.

Steerage horse hoe

Whether harrowed or not, it is of great importance to leave wheat land rough all winter, that is, with a round large clod upon the surface. These clods afford shelter from wind and frost to the young plants, and when gradually mouldered by frost, serve to increase the depth of the loose soil, and protect the roots of the plants from frost.

Autumn wheat is almost always sown broadcast, except, perhaps, in the neighbourhood of large towns, where it is sown with the *drill* (shown left); and the reason why the drill is used in that particular locality is the facility afforded by the drilled rows to hoe the land free of surface weeds, which invariably make their appearance where street manure is used. In England wheat is very generally drilled, for the reason just given. The method employed in weeding between closely drilled wheat plants is with the *steerage horse hoe* (illustrated above). This implement allows a slight deviation in the course of the horses to be rectified by the independent steering of the hoes so as to avoid destroying the juvenile wheat plants.

DIBBLING

There are other modes than those I have mentioned of sowing wheat on fallow ground. One of these is dibbling, and there are various ways of dibbling wheat. One is to make a hole not exceeding 2½ inches [6.5cm] deep, with a dibble not so thick as that used for planting potatoes in gardens, to drop one seed or two into it, and to cover them with earth with the foot; the holes being made 4 inches [10cm] apart, and 7 inches [18cm] wide between the rows. But this is a very slow process. A more expeditious plan is to use an implement made of a cylinder of wood 6 feet [1.8m] long, 4 inches [10cm] in diameter, and divided lengthways by the middle, to make into two dibbles. Pins of wood of a conical form, 3 inches [7.5cm] long, are driven perpendicularly 4 inches [10cm] apart into the apex of the curved side of the split cylinder. This implement forms a number of dibbles, by being laid along the ground with the pins downwards, which are pushed into the ground with the pressure of the foot, to make as many holes as there are pins. The implement being removed by means of a handle attached to its flat side, boys or girls drop two seeds into each hole, and cover them with earth. Another and more certain plan of dibbling is this. A flat, thin board of wood is provided with holes 4 inches [10cm] apart in the row, and the rows 7 inches [18cm] asunder. This is laid flat on the ground, when small dibbles are pushed through the holes to the requisite depth of 2½ inches [6.5cm] into the soil, the depth being determined by a shoulder on the dibble; two seeds are then dropped into the hole as each dibble is withdrawn; and when the board is lifted up from the space it occupies to another space in advance, the earth is brought over the holes and seed by the foot. It is asserted by those who have sown wheat by dibbling, that about one bushel is sufficient seed for one acre, and that the produce will be 5½ quarters per acre, that is, the produce bears a proportion of 44 to one of the seed sown.

SOWING USING THE PLOUGH

Another mode of sowing wheat has a similar effect in the appearance of the growing crop as ribbing with the small plough, and this is accomplished by using the *common plough* with a single horse, and depositing the seed and, along with it, if necessary, any species of manure dust, such as rape dust, in the furrow. The seed is dropped out of a *hopper* placed in the bosom of the plough, and the quantity is regulated by a grooved axle, made to revolve by a small wheel, which receives its motion by being carried along the ground with the plough.

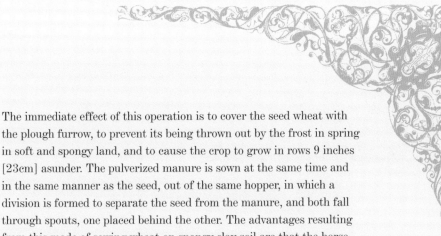

The immediate effect of this operation is to cover the seed wheat with the plough furrow, to prevent its being thrown out by the frost in spring in soft and spongy land, and to cause the crop to grow in rows 9 inches [23cm] asunder. The pulverized manure is sown at the same time and in the same manner as the seed, out of the same hopper, in which a division is formed to separate the seed from the manure, and both fall through spouts, one placed behind the other. The advantages resulting from this mode of sowing wheat on spongy clay soil are that the horse does not tread on the seed, and the seed requires no covering in by the harrow. It is necessary, however, to caution you in the use of rape dust and guano in contact with seed, as both are apt to affect the vitality of seeds, without the intervention of a little soil, or the previous mixture of a little earth.

Transplanting

Another mode recommended for cultivating wheat is transplanting. It is proposed to sow a small portion of ground with seed early in the season, and to take up the plants as they grow, divide them into single plants, and transplant them. By thus dividing the plants as they tiller into single plants at four periods of the season, a very small quantity of seed will supply as many plants as would cover a large extent of ground. Though wheat no doubt bears transplanting very well, yet as the scheme implies the use of much manual labour, it is questionable if it will repay the expense.

The object of these various modes of sowing wheat is the saving of seed – a great object, certainly, when it is borne in mind, that one fourteenth of the whole grain grown in the country is consigned to the earth in seed. From the statement just given in regard to the transplanting of wheat, it appears that 4 English pints [2.3 litres] of wheat are capable of supplying a sufficient number of plants for 1 acre. But practice has always tended against the use of a small quantity of seed; and the practice is sanctioned by the fact that though large quantities of seed are usually sown, in many seasons the young plants come up rather scanty.

Agricultural Wheel Carriages

I am induced to venture upon some practical considerations of the construction and application of wheel carriages.

Of the types of cart

Agricultural carriages are either four-wheeled waggons or two-wheeled carts. It is the the Scotch practice, with very few exceptions, that the two-wheeled cart only is used on the farm although a range of larger four-wheeled waggons can be found in England and these are used to haul vast quantities of hay, wheat in the sheaf and other commodities both around the farm and to market (shown left).

Though the cart, in general, is a vehicle very much diversified in structure to suit the numerous purposes to which, in a commercial country, it is applied; yet for the purposes of the farm its varieties lie within narrow limits, and may be classed under two principal kinds, the tilt, tip or coup *close-bodied cart*, and the close-bodied *dormant cart*; but these, again, vary as to size, forming single- and double-horse carts, which are merely varieties of the first. A third and less important kind is the corn or *hay cart*, used chiefly in the seasons of corn and hay harvest; and there are others not required on every farm, but are important to some, such as the *cage or cattle cart*, for carrying lambs and other livestock to market (opposite, top) and the water and liquid *manure carts*.

Farm waggon

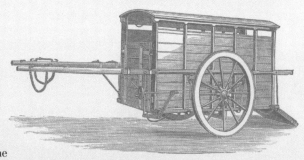

Cattle cart

TILT CART The tilt cart or tip cart (below right) is the most important vehicle of transport on the farm, and is employed for nine tenths of all the purposes of carriage required in the multifarious operations throughout the year. It is employed to convey manure of all kinds; to convey stone and other materials for draining and other operations; leading home turnips and potatoes, and for carrying produce of all kinds to market. For some of these operations the tilt cart is pre-eminently adapted, such as carrying and distributing of manures, or other matters that can be safely discharged by tilting.

The dormant cart, on the other hand, is sufficiently commodious when substances have to be carried that require to be discharged from the cart by lifting, such as grain in bags, and many other articles requiring to be conveyed to and from the farm.

Cart tipped

THE DORMANT-BODIED CART The dormant-bodied cart has its body constructed in all respects similar to that of the tilt cart; description of it is therefore unnecessary, except as regards the shafts. This cart requires no bolsters, but instead thereof the shafts are prolonged backward, taking the place of, and serving most of the purposes of, the bolsters; and at a slight glance, when viewed on the side, the cart has every appearance of a tilt.

TWO-HORSE AGRICULTURAL CART The two-horse agricultural cart differs only from the one-horse tilt, and its details, in being of larger dimensions, but especially in depth; the length is also increased a few inches, while the width remains nearly the same, and the limbers are stronger; but all the dimensions are variable, according to the tastes and objects of the owners.

CORN AND HAY WAGGON Next in importance to the carts already described comes the corn and hay wagon (shown overleaf), of which there are many varieties; but in many situations, and under certain

Hay waggon

systems of management, a substitute for these is adopted, in the application of the hay frame to the common close-bodied cart; and this, though somewhat injudicious in principle, is rather extensively adopted.

THE HAY FRAME The hay frame is a light rectangular piece of framework which consists, first, of two main bearers, which are fitted to lie across the cart, the one to the fore part, slightly notched upon the top rails, and leaning against the upper headrail, the top sides being at the same time removed; the other bearer is fitted in like manner to the back part, leaning against the door.

YOKING HORSES TO THE CART The harness required for workhorses in double and single carts has been already enumerated and described; and you will find a pretty good representation of a double horse cart on page 27. On placing that picture before you, you will more easily understand what is about to be said.

The shaft horse requires the following harness to be fully equipped: bridle, collar, haims, saddle and breeching. The use of the bridle, collar, and haims have been already described both inclusive; and these constitute the harness common to both plough and cart. The breeching is buckled to the back part of the wooden tree of the saddle, at such length of strap as suits the length of the horse's quarter. The saddle – as saddle and breeching together are commonly called – is placed on the horse's back immediately behind the shoulder, and strapped firmly on, in case of slipping off in the yoke, with the belly band; the breeching being put over the horse's hindquarter.

The back chain is fastened to the back chain hooks of the shafts of the cart, and there gets leave to remain constantly. The shafts are held up with their points elevated; the horse is told to turn and back under them, which he does very obediently, and even willingly; they are then brought down on each side of the horse; the back chain is adjusted along the groove of the saddle to such length as that the draught chains, when extended, shall be in a straight line to the axle; the shoulder clings, or draught chains, are linked to the draught hook of the cart, at such length as to be extended in the above line; the breeching chains are linked to the breeching hooks, of such length as to allow the breeching to hang easily upon the hams of the horse – not to chafe the hair – in his motion forward upon level ground, but so tight as that before the back chain hooks, with the back chain, slip as far back as they can upon the runner staple, the hams of the horse shall press against the band of the breeching sufficiently to keep the cart back, and to keep the cart back completely, before the horse's rump shall touch the front of the cart. The cart belly band is then buckled round the near shaft under the runner staple, just as tight as not to press against the horse's chest on level ground, and only when he goes uphill. All these adjustments of parts are made in a short time, with even a new horse, cart, or harness, and they require no alteration afterwards.

The harness of the trace horse is simple beyond the collar, haims and bridle, consisting only of two back bands, belly band and trace chains. The back band is placed where the saddle should be, and is fastened to the trace chains on either side with a buckle-like long-tongued hook. The trace chains are linked to the draught hook of the haims at one end, and fastened by a hook at the other end to a staple in the underside of the shafts; the point of which hook is always placed in the inside, to put it out of the way of taking hold of anything passing near the shafts of the cart.

The horses are now ready to start, in as far as the harness is concerned; but no cart, single or double, should be allowed to be used without double reins.

EGGS

The treatment which every kind of poultry should receive in autumn is that recommended in spring. Geese, ducks and turkeys should only hatch in spring and the early part of summer; for late hatchings of these never produce birds worth rearing, as they cannot attain in the same season a tolerable size and degree of fatness. But common fowls may be hatched throughout the summer, and even to a late period in autumn, and the chicks be reared to a useful state. What can constitute a more delicate dish than a chicken, boiled, roasted or broiled, at any season, but especially when the productions of the garden are in the highest state of perfection in summer and autumn?

Hens are fond of making their own nests and bringing out broods in cornfields and at the roots of hedges and shrubs; and when hens have their liberty during the day, it is impossible to prevent their following this inclination, which is common to all, under the best-regulated system; and so high a value do I set upon liberty to these creatures, on the score of health and strength of body, and flavour in flesh, that I would rather run the risk of losing a few broods in the year, by the fox and polecat making free with a self-set hen or two upon nests of their own seeking in cornfields, than see them cribbed in summer within a court of the largest dimensions. One of the daily cares of the henwife in summer and autumn is the gathering of eggs.

It should be the henwife's duty to visit every nest and collect the eggs from them every day, and the period of creating the least disturbance to the poultry in this duty is the afternoon, say between 2pm and 3pm, before the birds begin to retire to roost. A nest egg should be left in every nest; because it is an old established fact that all domesticated birds, at least, prefer to lay in nests containing eggs to those which are empty. Eggs are most conveniently collected in small hand baskets; and a short, light ladder will give easy access to all nests elevated above reach from the ground.

A wily fox creeps into the farmyard with evil intent

Whether eggs are retained for use at home, or disposed of to the egg merchant, it is of importance to keep them in a fresh state for some time. This end may be attained by preventing the air penetrating the pores of the shell, and the yoke coming in contact with its inside. A simple and effectual way of preventing the air penetrating the shell is to rub the egg over with butter when taken warm from the nest; and as simple a way of preventing the yoke adhering to the inside of the shell is to roll the egg from one side to the other every day. This treatment eggs should daily receive, whether kept for your own use or sold to the dealer; and it will preserve them in quite a fresh state for several weeks.

Hens begin to lay about the beginning of March, and continue to the beginning of October. They do not lay every day, that is, every 24 hours, some laying every other day, and some missing one day in three. They lay about two dozen of eggs at one period, then cease for two or three weeks, and again lay another two dozen, and so on for the number of months mentioned. Of all these months, however, they lay most constantly in March and April. After each period of laying they are inclined to sit on the eggs, and when it is not desirable for them to incubate, it is difficult, in most cases, to drive them from their propensity to clucking, as it is commonly called.

ONCLUSION

This subject brings me to the end of the farmer's year, the subject following being the ploughing of stubble, which I described at the beginning of winter, as the first operation of the year, and the first preparation for the next year's crop. In describing each operation as it fell to be executed in its own season, and in its proper relation to the one which preceded, and the one which was to follow it, I have endeavoured to make the description as minute and in as explicit terms as possible to enable you to execute it in the way it ought to be in the fields; or at least, to afford you data to compare any operation you see executed in the fields with the mode described here. Inasmuch, therefore, as the whole yearly operations of the farm have now been described, I might finish the work here; but there are still many other subjects, connected with every system of farming, upon which you require instruction, before you can be prepared to conduct a farm on your own account.

INDEX

acre (imperial) 247, 261, 265
Agricultural Improvement Society of Ireland 256
Agricultural Revolution 6
ammonia 118
animalized carbon 192
apple sauce 89

back band 170
back chain 296
baits 79
bailiff 18
Balmadies 11
bands 260
bandster 260
barley 100, *100*, *126*, 262, *266*
    boiled 113
    sowing 126
    stook *266*
    weeding 235
barn stool *96*
barrel churn 229
barrel churn 229–230
basket (for potatoes) 285
bathing can (or flask) *280*
bathing sheep *284*
bathing stool 283, *283*
beans 101, *101*, 271
    drill *137*, 138
    sowing 136-138
    within rotation system 137
bearing reins 46
beech nuts *175*
Belfast 256
Belgium 196
Bell, Rev. Patrick 268
Bell's reaping machine *268*
belly band 296
Bening, Simon of *15*
Berkshire boar *85*
biestings 115-116
binding, corn 265, 270
Birch or broom besom *79*
blackleg, (or quarter ill) 254
black soap 258
black-faced ram *282*
blacksmith 188, 201
blinders 46
blue vitriol 59
bone dust 192
Book of Hours 14
breasting knife *142*
breeches 216
breeching 296
bridle 46, 169, 170, 296

brine 76, 228, 231
brood sow 172
Brown, George 8, 10
brush *81*
buisting 281
bull calves 206
bull ridge 261
bullock holder 22, 208
bull's ring *207*
bushel (Imperial) *20*, 97, 98, 118, 130
butcher 188, 203–204
butter 227–230, 299
    beaters (or boards) *228*
    churn 224
    moulds 224
    salting 224
    weighing 224
    worker *228*, 228
Butter of Antimony 59
byre *68*, 70, 104, 204

cabbage 50–53, 195
calf's bladder 85
calves 69, 254
    stomach 231
    weaning 205–206
calving 110–117
candles 67
carding machines 77
carpenter 188, 201
carrots 50–53, 195
    sowing 196–197
carting 199, 166, 242, 247, 248, 265, 273, 285
carts 294–297
castration 213
cattle 21, 68–77, 203–211
    by-products of 76–77
    cart *295*
    droving 72–75
    feeding 68–71
cattle man 21, 70, 83, 94, 108, 209
chaff 95, 97
charcoal 258
charlock 235
charring 201
cheese 230–233
    cream 222
    mould (or chesset) 233
    press 224, 233, 233
    pressing 232
    Stilton, Gloucester, Wiltshire 222

    turning 233
    vats 224, 232–233
Cheviot ram *64*
chicken 298
Chippenham Cheese market *223*
Christmas 83, 88, 84
churning butter *227*, 227–230
clasp knife *92*
clay 290
clay 293
Clay 28–30
Clay cutters *28*
clearing stones 198–199
clipping floor 216
clover 130, 153, 188, 237
clucking 299
cochineal 67
cock 87
Cockieleekie soup 87
cocking 238–240
colic 79, 254
collar (for horse) 45, *45*, *46*, 170, 296
colls 240
colt 169–170
common draw hoe 195
compost 160–161
constructing a dam 213
cope stones 200
cordage 258
corn band 265
corn barrow *94*
corn basket *94*
corn cockle (or popple) 235
corn scoop *98*
coulter 36
cream 226–227
    jar *227*
    scallop (or skimmer) 224, *226*
creeping plume thistle 234–235
crop rotations 190, 238
cross ploughing 190
cross table 31
curd breaker 224
    cutter (or mill) 224, *232*, 232
curly greens 147
curry comb *81*
cutaneous eruptions 283

dairy 222–233
dairymaid 23, 83, 90, 108, 226
Davy, Humphrey 9
dead hedging 200
deep draining 42
deep drains *43*
deep ploughing (or sub-soiling) 246

301

Devon 102
dew 274
dibbling 291–292
dipping 284
ditcher's shovel 32, *32*
double horse cart *27*
doves 87
drafting ewes 280
drain scoops *44*
draining 42–44, 198, 246
draught chains 296
drover 60
drying flax 258
ducklings 185
ducks 87, 89, 185, 298
Duncan, Thomas 10
Dundee Academy 10
dung 190, 192, 196, 247
   hawk (or drag), 165
   spade *105*
   hill 84, 104–105, 158–159
Dutch hoe 235

East India Company 10
eaves *276*
Edinburgh, University of 10
eggs 91, 176–185, 298

fallow 189, 246–247, 290
farmer 18, 108
farrowing (or littering) 172–175
feering 37, *39*, 247, 248, 290
felts 258
fence steps *200*
fences, repairing 200
field gates 188, *202*
   fixing hanging post for *201*
   posts 201
fieldworkers 22, 193, 195, 220, 234, 240, 242, 286
firkins 224
flail 92
flauchter-spade *162*
flax 256–259
   sowing 134–135
   dodder (cuscuta europaea), 256
fleece, rolling 220
   weighing and packing *221*
fly-strike 280
foot picker *81*
fork *see* graip
fowl 86–91, 98, 176–185, 298
   feeding 89–91
   housing 89–91
   utility to man of *90*
fox 147, 152, 156, 298, *299*
French Burke, J 8
frost 195, 285, 289
frying pan (or lime) shovel *159*, 248

furnace 224
furrow slice *36*

gander 89
garden line 32
garters 216
gates 201
gathering, corn 265
Gauls 268
geese 88, 182–184, 298
gelding 254
Germany 196
gimmer 280–281
glue 77
gold of pleasure (*camelina sativa*), 256
Golden Age (The) 6
goose grease 112
goslings 183–184
graip (or fork) *70*, 158, 165, 278
granary 96
grass seed
   mixing 130
   sowing 130–133
   sowing machine 131–132
   types 130
gravel 28–30, 249
Great Exhibition 7
grieve 18
grouse 255
grubber (or cultivator) 163, 167, *167*, 190
gruel 62, 113, 175
guano 192, 293
gunpowder 258

haims 45, *46*, 296
haining 153, 213
hand draw hoe 234
hand hoe 192
hand pick *31*
hand stubble rake *265*
hare hunting 255
harness 170, 296
harnessing 45–49
Harrison and McGregor's self-raking reaper 269
harrowing 122–123, 132, 137, 163, 247, 248, 287, 290
harrows, English iron *123*
harvest 254–255, 259–271
hay, 199
hay knife *237*
hay rake, horse-drawn 241, *241*
hay rake 189, *239*, 238–239
hay stack
   building of 241–243, *242*
   round type 245
   thatching of 243–245
hay waggon *296*
haymaking 189, 237–245

hedges
   cutting down 141
   plashing 143–144
   pruning 139–141
   spade *32*, 235
   weeding 235–236
   weed hook 235
hedger 188, 200, 236
hedger's axe *145*
hedging 31–33, 139–145
hedging, dead 144, *145*
hemp 258
   sowing 135
henhouse 178
hens 87, 176–179, 298
henwife 298
hepatitis 254
Hereford bull *73*
High Farming 7, 11
Hillyard, Clark 8
hoeing 188, 194
hogs 258
hog's lard 112
Holland 196
horses 20, 45–49, 78–81, 241, 267–270, 273, 294–297
   breaking in 168
   -drawn mower *240*, 241
   hoe, (or scuffler) 194
   hoe, for turnips *192*, 192
   language addressed to 46–48
   stall *78*
hurdles (or flakes) *54*, 213

improved hand hoe *194*
Indian corn 184
Irish flax 256
iron hammer nut key *34*

James Caird 7
jars 224
Johnston, J.F.W. 9
Julius Work Calendar 14

kale 147
Keil 60
kemps 221

ladder *92*, 275
lambing 146–153
   assisting 149
   preparation for 147
   symptoms 148–149
lambs 280–81
   weaning 219–220
lantern 147
   for stable *80*
lard 85
Leicester sheep 219
lifting potatoes 285
lime 161, 290

liming 248
linseed 117
liquid manure 161
Liverpool salt 76
loam 28–30
Long-horned ox *75*
Loudon, J.C. 10, 101
Low, David 8, 100
Lussac, Guy 67

maiden ridge 260
mallet *55*
mane comb *81*
mangelwurzel 195
manure 104–105, 163, 190, 192, 197, 247, 287, 292
marble shelving 224
mason's hammer 249
mattock *32*, 32
Mechi, John Joseph 11
Michaelmas 184
milk
   cows 70
   dishes 222
   pail *115*
   pan *225*
   scalding 230
milking 114–115, 205
   stool *23*
Morocco leather 67
Morton, J.C. 10
mould board 36
mowing 93
   with a scythe *264*
mud-hoe (or harle or claut) *234*
muslin 224
mutton suet 67
   trade 60

nest box, for hens *177*
nest egg *298*
nievling 114

oats, 100, *100*, 125, 265
   sowing 124–125
   stooking *266*
   weeding 234–235
offal 76, 85
optical square *31*
oxen 204

Palladinus 268
parsnip 50–53, 197
partridge shooting 255
pasture 42, 188, 199, 211, 235, 283
   for cattle 204
   for ewes with lambs 153
pease 271
pickled ox-tongue 88
pigeon 87, 98, 185

pigs 82–85, 287
   by-products of 85
   bladders 85
   farrowing 172–175
   fattening 82–84
   sty 104, *174*
   trough 83
pint (English) 293
pint (Imperial) 231
pitchfork (or hand hay fork) 189, *238*, 238
pitching 201
Plaid (Shepherd's) 62
Pliny 268
plough 34, 170, 286–287, 292–293
plough reins 46
plough staff *34*
ploughing 34–41, 247
   effect on seed 128–129
ploughman 20, 78, 168, 188, 191, 237, 274
polecat 298
pole 31, *37*
pond 257
poppy, common red, smooth-headed 235
potato
   dunging 165–166
   handbasket *165*
   planting 162–167
   preparing the land for 162–163
   seed 164
   graip 285, *287*
   pits *288–289*
   plant *285*
   raiser *286*
   separator *286*
potatoes 285–289
poultry 86–91, 287, 298
   house *91*
poults 181
pruning knife 32
puddings 85
pullet 91
pulling flax 257

Quarterly Journal of Agriculture 11
quicklime 161, 248

rain gauge *18*
rake 260
Ransome's plough *35*
rape 188, 195
rape dust 292–293
rape oil 196
reaper knife-sharpener *267*
reaper-binder 269
reaping machine 267–269

rennet 231
ribbing 290
rick cloths *242*, 242
rick stand *93*
ricks 240
riddle 96, 98
ridge 41
roads 198
roller, cast iron *133*
rolling 199, 132–133, 247
rope 274, *274*
Royal Agricultural College 8
ruddle 60
rusky (or seed basket) 120
rye 101, *101*, 267
ryegrass 130, 153, 237

sack barrow *97*
saddle 296
salt and pepper 233
salting 76, 229, 231
sand 28–30
sandstone 249
sausages 85
sawdust 161
scales 229
Scotch drill plough *163*
scuncheon, laying of 250
scythe 237, 259, 263–264
   craddle *263*
   straight-snedded *261*
seed drill (for corn) *290*
seed germination 127
shade, for cattle 211
share 36
sheaf 260
shearling 282
shears *216*, 216
sheaves 92–95, 265, *266*, 273–4
shed 211
sheep 20, *62*, 254
   by-products of 65
   driving 60–63
   feeding turnips to 54
   fodder rack *56*
   foot rot 58
   mode of cutting up 65
   shearing *217*, *218*
   skin 66-67
   slaughtering 63–67
   tick (or keb or ked) 283
   utility to man of *66*
   washing 213–215, *215*
Shepherd 20, 62, 108, 146, 200, 209, 282–284
shepherd's crook 152, *152*
shepherd's dog 21, *155*, 154–157
shepherd's house *21*
Shire stallion *48*
shoulder clings 297
sickle *22*, 259, 261–263

singling 188, 192
slaking lime 248
soft soap 283
soil 28–30
sowing
    autumn wheat 290–293
    barley 126
    basket *121*, 121
    beans 136–138
    broadcast *122*, 291
    broadcast machine *132*
    carrots 196–197
    effect on seed 128–129
    flax, hemp 134–135
    grass seed 130–133
    machine for grass seed 131–132
    mangelwurzel 195
    oats 124–125
    parsnips 197
    rape 195
    sheet 120, *120*
    spring wheat 118–123
    turnips 191–192
spade *236*, *288*
spirit of tar 59, 62, 283
square-mouthed shovel *79*
stable 78–81, 104
stack 92, 274–275
    building of *275*
    trimmer *275*
stackyard 93, 255, 272, 275
staddle (or stathel) 273
stag (or Tom) turkey *180*
stallion *168*, 254
Steell, Gourlay 12
Steell, John 12
steeping flax 257
steerage horse hoe 291, *291*
stell, ancient *63*
    circular *60*
Stephens, Andrew 10
Stephens, Henry 10
steward 18, 93, 94, 96, 98, 108, 165
stone dykes 189
    building of 249–251, *250*
stoneware *222*, 225
stook 257, 265
stooking 266–267

stool (making a) 273
storing potatoes 288–289
straw 68, 102–103
    barley 102
    bean 103
    elevator *95*
    fork *83*, *103*
    oat 103
    rack *69*
    rye 103
    wheat 102
straw ropes 92, 94, 273
straw rope making 276–277, *277*
street manure 192
sulphur vivum 283
surface draining 43
swathes 239, 241
swedes 191, 195
swing trees 46, *47*, 48–49, 170
switching bill *140*

tallow 77
tanning 77
    bark 161
    nets 258
tedding 238, 241
    machine 241, *241*
thatching 278–279, *278*
thermometer *18*
thinning 197
thorn hedge 31, *140*
    breasting 142, *142*
thrashing 92–98
    machine 94–96, *93*, *95*
    rape 196–197
    ryegrass 245
throw crook (or wimble) *276*
tilt cart (or tip cart) *171*, 247, 295
tobacco 283
tobacco liquor 62
toothed sickle *259*
trace chains 45, 170–171
tramp pick 32
Transactions of the Highland and Agricultural Society 11
travis *69*
tup (or ram) 254, 282–283
turkey 87, 88, 180–181, 298
turnips 50–53, 70, 83, 147

cutter *146*, 147
drill harrows *236*
feeding sheep 54
hand hoe *23*
-lifter *50*
-picker *53*
method of pulling *51*
mode of feeding sheep *57*
mode of topping and tailing *52*
sowing machine 191, *191*
storing 52, *53*
trimming knife *51*
trough *70*, *147*
Tusser, Thomas 13

vernal grass 237
vetches 188
Victoria, Queen 6, 12
Virgil 14, 170
Von Liebig, Justus 9

waggons 294–296, *294*
water, for cattle 209
water furrowing 133
water scoop *211*
watering pool 209
wattling 200
weaning 117, 205–206
weather 27
Wedgwood-ware milk dish *222*
weed hook 234, *235*
weeding 197, 234–236, 256
weeds 247, 290
wheat 99, *99*, 254, 262
    apparatus for pickling *119*
    autumn sown 290–291
    pickling 118–119
    sowing in spring 118–123
    stook 267
    weeding 235
wheelbarrow *71*
whip 170
willow rod 244
wimble (or throw crook) 276, *276*
windrow 257, 239, 241
winnowing 96–98, *96*
wooden rule 32
wool 216–221, 281
worms 196